United States Marine Corps
Command and Staff College
Marine Corps University
2076 South Street
Marine Corps Combat Development Command
Quantico, Virginia 22134-5068

MASTER OF MILITARY STUDIES

SEA-BASED AIRPOWER – THE DECISIVE FACTOR IN EXPEDITIONARY OPERATIONS? NORWAY 1940 FALKLAND ISLANDS 1982

SUBMITTED IN PARTIAL FULFILLMENT OF THE REQUIREMENTS FOR THE DEGREE OF MASTER OF MILITARY STUDIES

AUTHOR: Major Willard A. Buhl
Academic Year 2001-2002

Mentor: Dr. Wray R. Johnson
Approved: _____
Date: _____

Mentor: Commander David A. Mee, USN
Approved: _____
Date: _____

REPORT DOCUMENTATION PAGE		Form Approved OMB No. 0704-0188

Public reporting burder for this collection of information is estibated to average 1 hour per response, including the time for reviewing instructions, searching existing data sources, gathering and maintaining the data needed, and completir and reviewing this collection of information. Send comments regarding this burden estimate or any other aspect of this collection of information, including suggestions for reducing this burder to Department of Defense, Washington Headquarters Services, Directorate for Information Operations and Reports (0704-0188), 1215 Jefferson Davis Highway, Suite 1204, Arlington, VA 22202-4302. Respondents should be aware that notwithstanding any other provision of law, no person shall be subject to any penalty for failing to comply with a collection of information if it does not display a currently valid OMB control number. PLEASE DO NOT RETURN YOUR FORM TO THE ABOVE ADDRESS

1. REPORT DATE (DD-MM-YYYY) 01-07-2002	2. REPORT TYPE Student research paper	3. DATES COVERED (FROM - TO) xx-xx-2002 to xx-xx-2002

4. TITLE AND SUBTITLE Sea-Based Airpower--The Decisive Factor in Expeditionary Operations? Norway 1940, Falkland Islands 1982 Unclassified	5a. CONTRACT NUMBER
	5b. GRANT NUMBER
	5c. PROGRAM ELEMENT NUMBER

6. AUTHOR(S) Buhl, Willard A. ;	5d. PROJECT NUMBER
	5e. TASK NUMBER
	5f. WORK UNIT NUMBER

7. PERFORMING ORGANIZATION NAME AND ADDRESS USMC Command and Staff College 2076 South Street MCCDC Quantico, VA22134-5068	8. PERFORMING ORGANIZATION REPORT NUMBER

9. SPONSORING/MONITORING AGENCY NAME AND ADDRESS USMC Command and Staff College 2076 South Street MCCDC Quantico, VA22134-5068	10. SPONSOR/MONITOR'S ACRONYM(S)
	11. SPONSOR/MONITOR'S REPORT NUMBER(S)

12. DISTRIBUTION/AVAILABILITY STATEMENT
APUBLIC RELEASE
,

13. SUPPLEMENTARY NOTES

14. ABSTRACT
See report.

15. SUBJECT TERMS

16. SECURITY CLASSIFICATION OF:			17. LIMITATION OF ABSTRACT Public Release	18. NUMBER OF PAGES 61	19. NAME OF RESPONSIBLE PERSON EM114, (blank) lfenster@dtic.mil
a. REPORT Unclassified	b. ABSTRACT Unclassified	c. THIS PAGE Unclassified			19b. TELEPHONE NUMBER International Area Code Area Code Telephone Number 703767-9007 DSN 427-9007

Standard Form 298 (Rev. 8-98)
Prescribed by ANSI Std Z39.18

REPORT DOCUMENTATION PAGE

FORM APPROVED - - - OMB NO. 0704-0188

1. AGENCY USE ONLY *(LEAVE BLANK)*	2. REPORT DATE	3. REPORT TYPE AND DATES COVERED *STUDENT RESEARCH PAPER*

4. TITLE AND SUBTITLE SEA-BASED AIRPOWER – THE DECISIVE FACTOR IN EXPEDITIONARY OPERATIONS? NORWAY – 1940, FALKLAND ISLANDS 1982	5. FUNDING NUMBERS *N/A*

6. AUTHOR(S) LIEUTENANT COLONEL WILLARD A. BUHL, USMC	

7. PERFORMING ORGANIZATION NAME AND ADDRESS *USMC COMMAND AND STAFF COLLEGE* *2076 SOUTH STREET, MCCDC, QUANTICO, VA 22134-5068*	8. PERFORMING ORGANIZATION REPORT NUMBER *NONE*

9. SPONSORING/MONITORING AGENCY NAME AND ADDRESS *SAME AS #7.*	10. SPONSORING/MONITORING AGENCY REPORT NUMBER: *NONE*

11. SUPPLEMENTARY NOTES *NONE*	

12A. DISTRIBUTION/AVAILABILITY STATEMENT *NO RESTRICTIONS*	12B. DISTRIBUTION CODE *N/A*

ABSTRACT: THIS ESSAY EXAMINES THE BRITISH SEA-BASED AVIATION IN SUPPORT OF TWO MODERN AMPHIBIOUS CAMPAIGNS: THE BRITISH CAMPAIGN IN NORWAY IN 1940 AND THE FALKLAND ISLANDS WAR IN 1982. EXPEDITIONARY MANEUVER WARFARE (EMW) OPERATIONS AND SUSTAINABLE LITTORAL POWER PROJECTION WILL REQUIRE VERSATILE AND FLEXIBLE SEA-BASED AIRPOWER TO ESTABLISH LOCAL AIR SUPERIORITY OVER THE FLEET. WHEREAS LAND-BASED AIRCRAFT CAN ATTACK FLEETS FROM GREAT DISTANCES, CURRENT LAND-BASED FIGHTERS CANNOT PROTECT THE FLEET FROM ATTACK WITHOUT EXTENDED AERIAL REFUELING. WHEN AIR SUPERIORITY CANNOT BE MAINTAINED OVER THE FLEET, AS EVIDENCED IN NORWAY AND THE FALKLANDS, NAVAL FORCES BECOME PREY TO LAND-BASED AIRPOWER. POWER PROJECTION FROM THE SEA OCCURS AS A RESULT OF MAINTAINING LOCAL AIR SUPERIORITY, NOT IN SPITE OF IT. UNTIL A SPACE-BASED SYSTEM CAPABLE OF PROVIDING AIR SUPERIORITY FOR SHIPS AT SEA IS FIELDED, "LEGACY PLATFORMS" SUCH AS AIRCRAFT CARRIERS AND THEIR EMBARKED AIR WINGS, DESPITE THEIR HIGH COST, WILL REMAIN ESSENTIAL NAVAL WARFARE PLATFORMS. FURTHER, HOWEVER, AND PERHAPS OF GREATEST IMPORTANCE TO THIS STUDY, AMPHIBIOUS FORCES ASHORE CANNOT RELY ON AIRCRAFT CARRIER (SEA-BASED) AIR SUPPORT IF A CREDIBLE LAND-BASED AIRCRAFT OR MISSILE THREAT TO SUPPORTING AIRCRAFT CARRIERS EXISTS.

14. SUBJECT TERMS (KEY WORDS ON WHICH TO PERFORM SEARCH) NORWAY 1940, FALKLANDS WAR 1982, BRITISH AIRCRAFT CARRIER OPERATONS IN SUPPORT OF MODERN AMPHIBIOUS CAMPAIGNS.	15. NUMBER OF PAGES: 49
	16. PRICE CODE: N/A

17. SECURITY CLASSIFICATION OF REPORT	18. SECURITY CLASSIFICATION OF THIS PAGE:	19. SECURITY CLASSIFICATION OF ABSTRACT	20. LIMITATION OF ABSTRACT
UNCLASSIFIED	*UNCLASSIFIED*	*UNCLASSIFIED*	

DISCLAIMER

THE OPINIONS AND CONCLUSIONS EXPRESSED HEREIN ARE THOSE OF THE INDIVIDUAL STUDENT AUTHOR AND DO NOT NECESSARILY REPRESENT THE VIEWS OF EITHER THE MARINE CORPS COMMAND AND STAFF COLLEGE OR ANY OTHER GOVERNMENTAL AGENCY. REFERENCES TO THIS STUDY SHOULD INCLUDE THE FOREGOING STATEMENT.

QUOTATION FROM, ABSTRACTION FROM, OR REPRODUCTION OF ALL OR ANY PART OF THIS DOCUMENT IS PERMITTED PROVIDED PROPER ACKNOWLEDGEMENT IS MADE.

EXECUTIVE SUMMARY

Title: SEA-BASED AIRPOWER – THE DECISIVE FACTOR IN EXPEDITIONARY OPERATIONS? (NORWAY, 1940; FALKLAND ISLANDS, 1982)

Author: Major Willard A. Buhl, U.S. Marine Corps

Thesis: This essay examines the British use of sea-based aviation in support of two modern amphibious campaigns: the British campaign in Norway in 1940 and in the Falkland Islands War in 1982. The purpose is to determine whether or not aircraft carriers (sea-based aviation) were at the root of the success or failure of British efforts.

Discussion: In April 1940, there were no airfields in central Norway capable of supporting modern, high performance aircraft. As the Norwegian campaign unfolded and the British faced a significant land-based air threat from the *Luftwaffe*, they failed to appreciate the tactical and operational potential of sea-based aviation. At the same time, British naval aircraft were technically inferior in design and capability compared to their *Luftwaffe* land-based counterparts in 1940. Nevertheless, despite determined attacks on British naval assets at the tactical level, at the operational level, the German command limited their campaign goals and did not exploit their advantage in the air to the extent possible. Their actions did, however, place great pressure on British sea based lines of communication in central Norway, the operational pivot of the campaign.

In 1982, against the Argentines, the British faced another opponent with superior land-based aviation. Although the British fully appreciated the need for air superiority, they employed a tactical scheme not unlike what had occurred in Norway. Nevertheless, the British were able to successfully contest the airspace above the Falklands and ultimately succeeded in defeating Argentine ground forces and ejecting them from the islands.

Conclusions: In Norway the British were able to execute successful counter-landings in northern and central Norway in April 1940. In the far north (Narvik area), where their forces could operate at the extreme range of German land-based airpower, the British achieved tactical success and periods of local air superiority. In central Norway (Trondheim area), the operational pivot of their campaign, they were unable to sustain their forces ashore which were "exposed to continuous and energetic German air operations."[1] The British logistical situation became desperate when the *Luftwaffe* destroyed the two principle ports supporting landing force operations in central Norway. The *Luftwaffe* enjoyed air supremacy in central Norway due to a complete absence of allied anti-aircraft weapons ashore and an Allied air defense that was limited to warships and British carrier-borne aircraft. The British carrier task force commander, Vice Admiral Wells, positioned his two carriers, *Ark Royal* and *Glorious,* 120 miles off the coast of Norway. Wells was not confident his embarked aircraft could sufficiently protect his carriers from German land-based aircraft, and ordered operational constraints to ensure the survivability of his aircraft carriers. With the British unable to contest the air, the ground forces ashore could not be sustained.

---

[1] Kurt Assman, *The German Campaign in Norway, Origin of the plan, execution of the operation, and measures against Allied counter-attack* . Tactical and Staff Duties Division (Foreign Documents Section), Naval Staff Admiralty. (Her Majesty's Stationary Office. 1948), 56. Assman was a Vice Admiral in the German Navy during World War II.

In the Falklands in 1982, the British again faced an enemy with numerically superior land-based airpower. Employing tactics similar to those used in the Norwegian campaign in 1940, the British task force commander, Vice Admiral John Woodward, positioned his two carriers, *Invincible* and *Hermes*, at the extreme eastern end of the tactical engagement area. The carriers' position allowed only 20-minutes time on station for the limited number of embarked Harriers. Although not optimal for combat air patrol over the task force and close air support for operations ashore, the British Harriers were able to gain and maintain air parity, and destroyed enough of their Argentine adversaries to enable successful landings and a successful campaign ashore.

Drawing upon the examples above, certain inferences can be drawn from which certain conclusions may be reached. Expeditionary Maneuver Warfare (EMW) operations and sustainable littoral power projection will require versatile and flexible sea-based airpower. Much as the British expeditionary forces faced land-based airpower threats in Norway in 1940 and the Falkland Islands in 1982, U.S. aircraft carriers will be required to establish local air superiority over the fleet. Whereas land-based aircraft can attack fleets from great distances, current land-based fighters cannot protect the fleet from attack without extended aerial refueling. When air superiority cannot be maintained over the fleet, as evidenced in Norway and the Falklands, naval forces become prey to land-based airpower. Power projection from the sea occurs as a result of maintaining local air superiority, not in spite of it. Naval aviation exists to provide air superiority for naval fleets and sealift assets. Power projection from the sea is possible in large part because of air superiority. Until a space-based system capable of providing air superiority for ships at sea is fielded, "legacy platforms" such as aircraft carriers and their embarked air wings, despite their high cost, will remain essential naval warfare platforms. Further, however, and perhaps of greatest importance to this study, amphibious forces ashore cannot rely on aircraft carrier (sea-based) air support if a credible land-based aircraft or missile threat to supporting aircraft carriers exists.

METHODOLOGY

This essay reviews the validity of sea-based airpower in support of a modern amphibious campaign in order to determine whether or not aircraft carriers are essential to the success of EMW.

Chapter one provides a case study on the British Campaign in Norway in 1940, focusing on the strategic background, operational planning, and the use of airpower.

      (1) The literature indicates that the British failed to grasp the tactical and operational capabilities of sea-based aviation in the Norwegian Campaign of 1940. Moreover, British naval aircraft like the *Swordfish* Biplane were technically inferior in design and capability compared to their German land-based counterparts.

      (2) However, despite determined attacks on British naval assets at the tactical level, at the operational level, the German command limited their campaign goals to support of tactical ground operations and destruction of port facilities.

      (3) Primary and secondary sources indicate that the German *Luftwaffe* established air superiority at the onset of the campaign in Norway but were on the verge of losing it in northern Norway before the withdrawal of British forces.

Chapter two provides a case study on the Falkland Islands War of 1982, focusing on operational planning and the use of airpower.

      (1) Primary sources indicate that sea-based airpower was an essential element of British naval operations during the Falklands Islands War.

      (2) Primary and secondary sources indicate that Great Britain won the Falklands War by destroying enough of the Argentine Air Force and Navy that further Argentine operations in the Falklands Islands were impossible. The Argentines were beaten even though large ground forces were still available on the mainland. To this end, airpower played a key role in the outcome.

Chapter three applies the lessons learned from the two case studies to current Marine Air Ground Task Force EMW.

Table of Contents

Introduction

The advent of the airplane meant that civil populations could no longer remain immune from the conflict, and that the two main elements of warfare, the sea and the land, were brought together by the unifying element of the air. In many respects, the implications of this change went unrecognized in 1918, leading to developments in military thinking and equipment in the inter-war years which left those who were destined to pick-up again the cudgels in 1939 sadly ill prepared for their task. But the learning process for both sides was soon to be rekindled, for their respective campaigns in Norway in 1940 exemplified a new truth – that land, sea and air forces would henceforth have to operate in co-operation and harmony with each other, a failure in one could have a decisive impact on the efficiency and capabilities of the others... The era of joint warfare had arrived.

> *- Air Commodore Maurice Harvey*
> *Scandinavian Misadventure*

Naval aviation exists to provide air superiority for naval fleets and sealift assets. Power projection from the sea is facilitated by air superiority. Until a space-based system capable of providing air superiority for ships at sea is fielded, aircraft carriers and their embarked air wings, despite their high costs, will remain essential naval warfare platforms. Aircraft carriers exist to gain and maintain air superiority over the fleet. Whereas land-based bombers can attack fleets from great distances, current land-based fighters cannot protect the fleet from attack without extended aerial refueling support. But when air superiority cannot be attained over the fleet, as evidenced in Norway and the Falklands, naval forces become prey to land-based aviation. Today's anti-ship missiles, with 90 nautical mile ranges or better add an additional threat to naval forces. Power projection from the sea, however, is primarily rooted in the ability of sea-based aviation to establish and maintain air superiority.

In light of the above, was sea-based airpower the decisive factor in the British failure of the Norwegian Campaign in 1940? And what of the success of the Falklands War of 1982? The purpose of this study is to assess British use of sea-based airpower in the Norwegian Campaign of 1940 and the Falklands War of 1982 in terms of the lessons that can be applied to U.S. sea-based operations of the future. The purpose is to determine the relationship between sea-based airpower and the failure or success of the British expeditionary operations. From this analysis, certain inferences can be drawn, from which certain conclusions may be reached with respect to current Marine Air Ground Task Force (MAGTF) Expeditionary Maneuver Warfare operations.

Chapter I
The Norwegian Campaign 1940

Unlike their adversaries, the Allies had no clear strategy or aim in Norway in 1940. The Allied

Expeditionary Force was en-route to Norway to conduct what was planned as an unopposed

occupation of key ports in order to deter potential German aggression in Scandinavia.[2] By

occupying Norway, the Allies also hoped to deny valuable raw materials, especially Swedish iron

ore, to the German armament industry. The Germans had other plans, however, and were already in

the process of executing *Fall Weseruebung,* their plan for the invasion of Norway.

Fall Weseruebung was an extremely daring undertaking in that "it was the first occasion in the

history of warfare that the three Services had participated in a tightly controlled and coordinated

joint exercise in which failure of any one of the component parts would have irrevocably damaged

the whole operation."[3] The Germans landed isolated units by air and sea along 1,200 miles of

Norwegian coastline, widely separated in space and in some cases time. These landings were

achieved in the face of the vastly superior British Royal Navy during a period of highly

unpredictable winter weather conditions in Norway. Emphasizing speed, surprise, and airpower in a

multi-axis attack, the Germans established a milestone in the history of joint warfare, by the

effective projection of their military forces thousands of miles from their bases in Germany.

The Allies slowly and inadequately reacted to the German invasion. The British operation was

equally joint; however, there was little cooperation with the Norwegians and even less between the

British Services. This was especially true between the British Army and the Royal Navy. British

command and control was lamentable, and British soldiers were ill prepared and equipped to fight.

More importantly to the outcome of the campaign, perhaps, was that British aircraft, for a number of

---

[2] Author's note, the Allied Expeditionary Force was predominantly British but included French naval vessels and ground forces and Polish ground forces. See for example

[3] Maurice Harvey, *Scandinavian Misadventuer, the Campaign in Norway 1940,* (Tunbridge Wells, United Kingdom,. 1990), 36.

reasons predominantly related to pre-war resource decisions, were in short supply and many of the types employed were obsolescent.

Strategic Considerations

The German idea of securing naval bases in Norway in a war with Britain originated decades before World War II.[4] Based on their experience in World War I, the German Navy (*Kriegsmarine*) was fixated on breaking out of the confines of the North Sea to carry the war to their anticipated adversaries in the Atlantic. After the successful campaign in Poland, the German Army Chief of Staff, General Halder, expressed doubts that the planned campaign in France, *Fall Gelb,* would result in a rapid occupation of French Atlantic ports. Consequently, German naval planners looked for an alternative and began to plan for action in Norway.[5]

Grand Admiral Raeder, commander of the *Kriegsmarine*, brought Norway's strategic significance to Hitler's attention on 10 October 1939. A vigorous proponent of German military action to occupy Norway, he outlined several key points. First, it was essential to prevent Britain from occupying Norway to keep Sweden away from British influence. Second, possession of Norway would protect German sea lines of communication (SLOCs) in the Baltic. Third, occupation of Norway would prevent the British from interrupting the supply of Swedish ore to Germany. Finally, securing Norway would deny Britain from using bases there to expand the air war over Germany. Raeder also told Hitler that Britain could disregard Norwegian neutrality and interrupt ore traffic from Narvik to German ports without occupying Norway. Most importantly, Raeder presented the German Naval Staff's conclusion that "the maintenance of German naval supremacy in the Baltic and the continued

---

[4] Gerhard Weinberg, *A World at Arms*, (Cambridge University Press, United Kingdom 1997), 113. Many officers in the German Navy were influenced by the writings of Vice Admiral Wolfgang Wegener, who argued in a book published in 1929, *Die Strategie des Weltkrieges*, that the main function of the German Navy in World War I was to keep the sea lanes open for German merchant shipping, but that this could never be accomplished from "the dead angle of a dead sea" as he described the German and Danish North Sea coast line. See also Telford Taylor's *March of Conquest*, (Nautical and Aviation Publishing Company, New York, 1991), 83.
[5] Weinberg, 113.

supply of Swedish ore were both vital to Germany's conduct of the war" and "the loss of Norway to England would be synonymous with losing the war."[6]

Other principle members of the German Naval Staff, indeed Hitler himself, remained unconvinced about the need to take military action against Norway in 1939. While occupation of Norway offered an operational base for naval and air activity against Britain, the Naval Staff recognized that the German center of gravity for naval warfare against Britain was the Atlantic.[7] Norway was far from the Atlantic and separated by the Greenland-Iceland-United Kingdom (GIUK) gap. Additionally, Russia and Germany were at peace and the Arctic did not hold any strategic significance for Germany at that time.[8]

British maritime strategy had not changed since World War I. The British naval barrier between the Orkney Islands and southern Norway denied access to German surface vessels into the open seas and also equally cut Germany from her lifeline of vital imports. This economic blockade played a major role in weakening German economic strength and resolve during World War I.[9] Indeed, the British Chiefs of Staff had prepared an assessment for Britain's Foreign Secretary, Lord Halifax, detailing how Britain should deal with Norway in the event of a German attack on that country.[10] This assessment concluded that Germany's interest lay in maintaining a neutral Norway unless the latter adopted such a sympathetic attitude towards the Allies that the iron ore supplies were threatened. In addition, British naval doctrine required the establishment of sea control as a prerequisite for amphibious assault operations. The report dismissed as impracticable the possibility of German amphibious operations against Norway's western seaboard -- a miscalculation of critical importance."[11]

---

[6] Assman, 1. See also Weinberg, 113, and 252 regarding Admiral Raeder's attempts to deflect blame at the Nuremburg War Trials for his personal instigation and involvement influencing Hitler's decision to invade Norway.
[7] Ibid., 1.
[8] Ibid., 1.
[9] Harvey, 37.
[10] Thomas Kingston Derry, *Campaign in Norway* (Her Majesty's Stationary Office, London 1952), 21.
[11] Harvey, 13.

Sir Winston Churchill reentered the Admiralty on 3 September 1939 and immediately advocated action to guard Britain's northern flank.[12] He soon learned that experts from the Ministry of Economic Warfare and the Admiralty had noted that, out of the 11 million tons of iron ore used by Germany, 9 million tons were imported from northern Sweden. In the summer the iron ore was shipped from the Swedish port of Lulea, in the Gulf of Bothnia. In the winter, when the Gulf of Bothnia was frozen, the ore came via Narvik on the west coast of Norway. As Churchill stated in his memoirs, "To respect the corridor would be to allow the whole of this traffic to proceed under the shield of neutrality in the face of our superior sea-power. The Admiralty Staff were seriously perturbed at this important advantage being presented to Germany, and at the earliest opportunity I raised the issue in the Cabinet."[13]

Realizing that such imports were vital to Germany's armament industry, the ensuing debate centered around three main options: 1) The Royal Navy could mine the Leads, Norway's coastal approaches, to force merchant traffic to the open seas; 2) Britain could send surface ships into Norwegian waters for the same purpose; or 3) Britain could use military means to obtain control of the port of Narvik and, potentially, the iron ore workings themselves.[14] These options were discussed inconclusively in the months following the outbreak of the war, with the British and French leadership strongly influenced by the fact that merchant ships refused to sail during September.[15]

British and French attention was refocused on Scandinavia as a result of the Soviet Union's invasion of Finland on 30 November 1939. The Soviet Union's ruthless and unprovoked attack on her smaller neighbor produced widespread international sympathy for the Finnish cause, yet there was little that the British and French could do for this distant and remote country.[16] At a meeting of the Supreme War Council in Paris on 19 December, the French produced "extracts from a memorandum

---

[12] Ibid., 13.
[13] Winston Churchill, *The Second World War* Volume I, (Cassell, London 1948), 478.
[14] Harvey, 14.
[15] Ibid., 14.

by the eminent German Industrialist Fritz Thyssen, who maintained that Swedish iron ore was of critical importance for the German armaments industry."[17]

At this time the French Premier, Daladier, was under attack in Parliament, by the press, and in the cafés for the conspicuous inertia of his war policy. Anxious to deflect German interest away from France and the Low Countries, as well as relieve domestic pressure, the French pressed their British Allies to take action in Scandinavia. The British War Cabinet strongly considered action and on 20 December 1939 the Military Coordination Committee, asked the chiefs of staff to consider dispatching troops to Narvik and northern Sweden.[18]

German attention became more focused upon Norway as they realized that the British were contemplating occupation in the wake of the Finnish invasion. Hitler subsequently directed the Armed Forces High Command (*Oberkommando der Wehrmacht-OKW*) to prepare a draft plan, *Studie Nord,* for the military occupation of Norway.[19] Hitler's decision to begin planning for offensive operations in Norway was also influenced by German intelligence that the Allies were planning action there in the wake of the Russian invasion of Finland, which had begun the previous month.[20]

In February 1940, the French and British Governments directed their military staffs to create detailed plans for a landing in Norway in the third week of March.[21] Details of Allied intentions were leaked to the Scandinavian press, hardening attitudes of the Swedes and Norwegians against Allied intervention. This revelation also increased German military and political attention regarding Norway. While the balance of opinion throughout the German Navy continued to favor maintaining Norway's neutrality, German concerns came to the fore when the German auxiliary ship, *Altmark,* with 299

---

[16] Martti Haikio, *Scandinavia during the Second World War, The Race for Northern Europe, September 1939-June 1940* (University of Minnesota Press Minneapolis 1983), 72.

[17] Francois Kersaudy, *Norway 1940*, (Collins and Sons, London, 1990), 19. Harvey also states that Thyssen even went as far as to suggest that the war would be won by the country which secured control of the Swedish ore (Harvey, 16.).

[18] Haikio, 73.

[19] Earl Frederick Ziemke, *The German Northern Theater of Operations* (Washington D.C.: Center of Military History, 1989), 9.

[20] Assman, 6-7. See also Harvey, 41.

British prisoners aboard, was boarded by the British destroyer *Cossack* in Norwegian territorial waters on 16 February.[22] "The deliberate action of the *Cossack* convinced Hitler [that] the British no longer intended to respect Norwegian neutrality and, on 19 February, he demanded a speed-up in the planning for *Weseruebung*."[23]

Operational Planning

Buoyed by his victory in Poland the previous summer, "Hitler was able to dominate his senior military commanders to an extent which assured fixity of purpose which was quite lacking on the other side of the North Sea."[24] For the "first time direct control of operational planning was taken out of the hands of service commands and vested in Hitler's personal staff, the OKW." [25] Led by Captain Theodore Kranke, Commanding Officer of the cruiser *Scheer*, the staff for *Weseruebung* was formed by Hitler's direction in February 1940 as a special section of the Operations Staff of the OKW. The emphasis on a planning staff led by naval planners is due, in part, to the fact that the Army and *Luftwaffe* were immersed in preparations for *Fall Gelb*, the upcoming German invasion of France and the Low Countries. Additionally, the *Kriegsmarine* had conducted the initial planning for *Weseruebung*.

---

[21] Harvey, 21.

[22] Kersaudy, 24.

[23] Ziemke, 16.

[24] Harvey, 41. See also General Siegfried Westphal's *The German Army in the West*, (Cassell and Company, London 1951), 77. General Westphal provides a cogent description of Hitler's demands of "absolute concurrence with his ideas and unquestioning obedience" on the part of assembled army leadership on 23 November 1939 at the Reich's Chancery.

[25] Ziemke, 14. There have been a number of explanations as to why Hitler broke precedent and assigned the OKW to plan an operation. Ziemke suggests that the *Luftwaffe* was not put in charge as a result of a recent security breach by an *Luftwaffe* Major who inadvertently landed behind Allied lines in Belgium with plans for the upcoming campaign in France, *Fall Gelb*. Some authors have suggested that Hitler may also have lost confidence in the Army as a result of delays in execution of *Fall Gelb*. Other historians suggest that Hitler was convinced that this joint operation was so different from anything previously executed that it ought to be planned by the OKW, not only in broad outline (normal procedure was to issue broad outline and task a Service to do the detailed planning) but in the details based upon four reasons. First, the size and difficulty of the operation meant that any one service was not qualified to direct the other two. Second, the operation required close collaboration with the Reich's Ministry of Foreign Affairs. Third, was the need for improved secrecy. Finally, *Fall Gelb* was the Army Staff's main effort and had their focus. For an interesting primary source insight into Hitler's relationship with the Army, see General Westphal, Chapter III, The Army Is Stripped of Power, pp.52-3.

Building upon *Studie Nord*, Kranke's staff focused operational planning by dividing Norway into six strategically important areas, which contained most of Norway's population, industry, and trade. Assuming that control of these areas would enable the Germans to control the entire country, Kranke's staff proposed to execute simultaneous landings at the key ports of Oslo, Kristiansand, Arendal, Stavanger, Bergen, Trondheim, and Narvik. The simultaneous capture of these ports was expected to achieve a *coup de main*, preventing the Norwegians from mobilizing eight of their estimated sixteen regiments, nearly all of their artillery, and would deny them use of almost all of their airfields.[26] Perhaps of greater importance, the Germans assumed that seizure of these ports would also forestall potential Allied intervention.

Demonstrating unprecedented joint innovation in order to achieve maximum operational surprise and synchronization, *Weseruebung's* planners directed half of the ground forces designated to land on "*Weser* Day" be transported by sea in fast warships, and the other half by JU-52 transport aircraft. Additional key points were also selected in Denmark, to be captured primarily by land forces from across the German border.[27]

Speed and surprise, enabling swift capture of all initial objectives, followed by a rapid build up of follow-on forces by air and sea lift, was essential to German success in *Weseruebung*.[28] An army corps composed of the 22nd Infantry Division (airborne), the 11th Motorized Rifle Brigade, one mountain division, and six reinforced infantry regiments was task organized to accomplish the

---

[26] Ziemke, 15. Physical characteristics were major planning considerations. Norway is extremely mountainous with lines of communication formed by river valleys. In 1940, 85% of the national population and 100% of the industry were located in the southern cities of Oslo and Trondheim and in the valley that connects them.

[27] Ibid., 37. Denmark was included in the planning because the *Luftwaffe* argued that they needed Aalborg, on the Jutland peninsula, in order to support the campaign and assist in the isolation of Norway through aerial attacks against British naval forces. General Kessler states that poor signal communications in Oslo required the bulk of the *Luftwaffe* to operate from bases in Germany (Schleswig, Westerland, Fuhlsbeuttel, Uetersen) and Denmark (Aalborg, Copenhagen, Skagen) throughout the early days of the campaign. See also Kessler, Ulrich O. (former General der Flieger German Air Force W.W.II and Chief of Staff, Fliegercorps X during Operation *Weseruebung*), *The Role of the Luftwaffe in the Campaign in Norway 1940* (Army Historical Center Monograph, 1954), 10.

[28] Assman, 8. "The general operation comprised two separate and independent phases: a) Sudden occupation. b) Reinforcements in troops and equipment for enlarging the initial positions." See also Derry, 35.

operation.[29] In addition to the use of nearly the entire *Kriegsmarine*, ground forces were supported by

over 1,000 aircraft from *Fliegerkorps X.*[30] Another unprecedented German innovation included the

use of three companies of *Luftwaffe* paratroopers, or Fallschirmjaeger. The paratroopers, an essential

element of the surprise hoped for on the part of *Weseruebung's* planners, were assigned airfield

seizure missions in order to ensure the smooth introduction of air-landed follow-on forces.[31]

The capture of two decisive points -- the Norwegian Capital of Oslo and the Danish Capital of

Copenhagen -- was expected to paralyze the Norwegian and Danish Governments. The Germans

identified these governments as the center of gravity for the operation. The early capture of the royal

families of Norway and Denmark was also an essential element of the German strategy to effect a

quick surrender and political settlement.[32]

General der Infanterie von Falkenhorst, a corps commander and veteran of World War I service in

Finland, was interviewed and selected by Hitler to command Operation *Weseruebung* on 2 February.

Falkenhorst was directed to complete planning with his XXI Corps Staff (subsequently known as

Group XXI) with the objectives of securing the ports and imposing "such firm control on the country

that resistance or collaboration with the British would prove impossible."[33]

The German invasion of Norway was "a seaborne pounce, based on the correct guess that the far

---

[29] Ziemke, 15.

[30] Ibid., 56. Author's note - a lecturer at the US Marine Corps Command and Staff College has referred to the German JU-52 transport aircraft as "perhaps the real secret weapon of the Blitzkrieg." In the early weeks of *Weseruebung,* 582 transport aircraft moved 21 battalions, nine division and regimental staffs, and a number of mountain artillery batteries, plus naval and air force personnel and equipment. It was estimated that the air transport units flew 13,018 missions, carrying a total of 29,280 men and 2,376 tons of supplies. See also Kessler, 7- 9. The *Luftwaffe* assigned Colonel D.R. Freiherr von Gablenz, the Director of the *Deutsche Lufthansa*, to Flieger Korps X headquarters to organize and oversee air transportation and logistical support for the operation.

[31] Kessler, 6. General Kessler was of the opinion "that the display of air power as shown during the landing of troops and material with hundreds of planes roaring over southern Norway helped much to facilitate the occupation of the first objectives."

[32] Kersaudy, 49. The Germans hoped to occupy Norway without fighting, expecting "the Norwegian armed forces to show neither the desire or the ability to offer effective resistance." It was thought that the German position could be consolidated by diplomatic means, with the Norwegian Government initially being assured of "as much independence as possible" in internal affairs. Kersaudy's work offers an excellent study of the political aspects of the Norwegian campaign. See also Ziemke, 15.

[33] Harvey, 44.

stronger British and French navies would never be able to react in time."[34] Seizure of the main

objectives and securing use of the airfields by the *Luftwaffe* would allow the conquest of Norway to

be conducted without the threat of British land-based airpower operating over the Norway (due to

range limitations from Great Britain to Norway). Thus, the British would be forced to rely on sea-

based airpower from carriers, if at all.

Role and Effectiveness of Airpower

*"The campaigns in Poland, Holland, Belgium and France, and last but not least, in Norway had
proved unequivocally how important air supremacy is in modern war."*

General der Flieger Karl Koller, 1945

German surprise was achieved on 9 April in every port except Oslo. This, despite the fact that the

operation was nearly compromised on 8 April when the Polish submarine *Orzel* sank one of the ships

from the 1st Sea Transport Echelon, the *Rio De Janiero*, and her German survivors (in uniform) were

picked up by Norwegian authorities.[35] Denmark was overrun in one day, allowing important airfields

in Jutland to be used as forward air bases in support of Norwegian operations. With the exception of

the sinking of the German cruiser *Bluecher* by a coastal defense fort at Oslo, the initial landings were

successful and met little Norwegian resistance once established on the shore.[36] The *Bluecher's*

sinking, however, delayed the capture of Oslo by half a day and enabled the King and principle

---

[34] Richard Humble, *Aircraft Carriers*, (Winchmore Publishing Ltd, Hadley Wood, England, 1982), 52.

[35] Assman, 21.

[36] Kessler, 6. Initial paratroop landings at Oslo's commercial airport, Fornebu, were canceled by bad weather and anti-aircraft fire. The airfield seizure was successfully conducted (three hours behind schedule) at 0830 when 3,000 infantry were air landed by JU-52 transport aircraft over a two-hour period. Initial aircraft were fired upon, resulting in casualties on both sides that included 31crashed German aircraft. The paratroopers turned back on orders from the *Luftwaffe* chain of command. The second waves were predominantly composed of army infantry battalions. They followed Army orders to execute the landing despite unfavorable landing conditions and opposition forces on the ground. See also Ziemke, 52, or Cajus Bekker's *The Luftwaffe War Diaries*, (Ballantine Books, N.Y. 1969), 102-9. Bekker provides a riveting account of the tactical actions at Fornebu, relating that the German transport commanders believed that the *Fliegerkorps X* recall order was a Norwegian deception ploy and chose to disregard orders and execute the landing despite adverse weather and apparent enemy resistance.

members of the Norwegian Government to flee the city with 50 tons of gold from the Norwegian Treasury.[37]

The German Naval Staff had decided to allow the 1st Sea Transport Echelon a week's advance sailing in order to reduce the risk of discovery before the assault. The effects of this decision on the operational synchronization of logistics were apparent when oil tanker support failed to arrive in Narvik immediately following the landings.[38] The resulting delay enabled British naval forces, including the battleship HMS *Warspite*, to respond, trap, and destroy all of the German destroyers in the fjords surrounding Narvik. German naval losses at Narvik totaled 50% of all destroyers in the *Kriegsmarine*.[39]

The British made two crucial decisions that provided unwitting assistance to the German execution of *Weseruebung*. First, on 8 April, they recalled their mine laying ships well clear of the coastline to rejoin their covering force (battle cruiser *Renown* and light cruiser *Penelope*). This effectively removed these ships from a position to discover and engage German naval forces transiting the coast of Norway. Second, and more importantly, the Admiralty commanded that, given the strength of the German force at sea, every ship was needed for fleet operations. This effectively delayed the British *Plan-R.4*: the landing of British and French troops in Norway. [40] Thus, the Admiralty's measures, adopted to secure the traditional object of a decisive encounter at sea, failed and actually deprived the British of their best chance to restore the position on land.[41]

---

[37] Ziemke, 51. In addition to the immediate impact of (government led) continued resistance against the German invaders, this setback was to have strategic implications as the King ultimately fled to Britain and established a government in exile. The King's government in exile was a symbol of Norwegian resistance throughout the remainder of the war. See also Harvey, 305. Author's note - Ironically, the Oscarsborg Fortress's 280mm guns that sunk Germany's newest cruiser, *Bluecher,* were German (Krupp model 1905), and the torpedoes were of Austrian manufacture. The fort was built during the Crimean War in the previous century.

[38] Assman, 13. Only one of the five scheduled tankers arrived, creating a fatal delay in refueling.

[39] Ibid., 44. It can be argued, however, that the resulting 2,500 sailors stranded ashore reinforced General Dietl, the ground commander at Narvik, with sufficient strength to delay the Allies' only successful ground effort in Norway and thereby made an operational contribution to the overall success of *Weseruebung*.

[40] Derry, 26: "No expedition should be risked until the naval situation was cleared up."

[41] Ibid., 26.

While possession of Oslo was the key to the German occupation of Norway, the city of Trondheim was the operational pivot point of all German military operations north of Oslo.[42] Initial Allied strategic and operational planning emphasis had consistently focused on the occupation of Narvik, in order to deny iron ore shipments to Germany.[43] As the campaign unfolded, however, British military and civilian leadership realized Trondheim's importance and adjusted their military planning accordingly.

The Allies executed brigade strength counter-landings between 12 and 18 April at Namsos and Andalasness, ports to the immediate north and south of Trondheim (respectively). These landings forced the German commander at Trondheim to conduct holding operations while awaiting reinforcement from forces advancing north from Oslo.[44] In an effort to cut British and Norwegian lines of communication back to the coast, the Germans dropped a parachute company on the town of Dombass, an important railroad junction in the Gudbrandsdal Valley between Oslo and the port town of Andalasnes. But the operation was not coordinated between the *Luftwaffe* and the Army resulting in encirclement by the British and the capture of the German paratroopers. [45]

German land-based airpower concentrated on interdiction efforts, especially Allied shipping and ports of entry in the Trondheim area of operations. Admiral Sir C. Forbes, the British Commander in Chief of Home Fleet, had previously expressed grave concerns about conducting naval operations within range of German land based aircraft commenting, "I do not consider Operation [Trondheim] feasible unless you are prepared to face very heavy losses in troops and transports."[46]

---

[42] Assman, 53. See also Ziemke, 77.

[43] Harvey, 102. Trondheim was the medieval capital of Norway, the third largest city, and an important agricultural and industrial area. It was also a major rail center and terminus from Oslo and possessed a large deep-water anchorage with extensive quays and dock facilities. Its recapture would not only provide a secure political and strategic base for any future operations in the south, it would also effectively isolate German forces in the north.

[44] Ziemke, 78-80.

[45] Harvey, 129-30. Harvey comments, "Whilst a spirited initiative, the paratroop landing was unlikely to have been successful unless it could be quickly supported by an infantry breakthrough in the Gudbrandsdal."

[46] Ibid., 104.

These fears were realized when German aircraft conducted seven hours of determined attacks against the British cruiser *Suffolk* off the Norwegian port of Stavanger, causing her to return to Britain with the "sea lapping at her quarterdeck."[47] Later, German air raids easily destroyed the towns of Namsos and Andalasness and their vital port facilities (all constructed primarily from wood). Unable to gain even local air superiority and with severed lines of communication (logistics), the Allies abandoned their efforts to seize Trondheim and central Norway.[48] Thus, "while the German Air Force was not able to carry out its strategic mission, to the extent of preventing enemy landings in Norway, it was operationally and tactically effective in keeping the Allies from establishing secure bases, and contributed greatly toward forcing their subsequent withdrawal."[49]

While German airpower and rapidly advancing ground forces from Oslo were instrumental in reducing Trondheim, the German forces at Narvik had maintained a very tenuous hold on the town. The Germans' loss of sea control and their extended aerial line of communication enabled Allied forces (British, French, Norwegian, and Polish) to isolate them. Hitler was quite alarmed about the situation in Narvik and attempted to intervene personally in the tactical conduct of operations there.[50] Generals Keitel (Chief of the OKW) and Jodl (Chief of Operations, OKW) were able to assuage Hitler, permitting General Dietl, the resourceful German commander at Narvik, the opportunity to conduct a skillful and stubborn delaying action. General Dietl was ultimately forced to abandon Narvik on 28 May, but continued to delay east toward Sweden along the Lapland Railroad line connecting Narvik to the Swedish ore fields. Timely reinforcement by air dropped paratroop units

---

[47] Assman, 54. The *Suffolk* was beached at Scapa Flow to prevent her from sinking.

[48] Kessler, 11.

[49] Ziemke, 111. Also, for an excellent summary of scholarly works addressing the impact of German airpower in *Weseruebung*, see Jack Adams, *The Doomed Expedition, The Campaign in Norway 1940*, (Leo Cooper Ltd., London, 1989), 169-77 and Adam Claasen's *Hitler's Northern War, The Luftwaffe's Ill-Fated Campaign, 1940-1945*, (University Press of Kansas, 2001).

[50] Telford Taylor, *The March of Conquest*, (Simon and Shuster, New York.1991), 138-41. Taylor titles this section, "The Fuehrer's Crisis of Nerves", describing different times during *Weseruebung* when Hitler became involved in operational and tactical command and control decision-making. His assessment is supported by many other primary

prevented Allied forces from overwhelming his command. General Dietl was thus able to continue his stubborn resistance until the Allies withdrew from Narvik on 8 June.[51]

When Germany invaded France and the Low Countries on 10 May 1940, the Allies decided to withdraw their forces from Norway. In his wartime memoirs, Churchill stated, "On May 24, in the crisis of shattering defeat, it was decided, with almost universal agreement, that we must concentrate all we had in France and home."[52]

Any attempt to analyze the ineffectiveness of British airpower in Norway in 1940 must be prefaced by an examination of the context in which it was employed. The British employed airpower in Norway primarily in the form of maritime reconnaissance, along with long-range bomber operations from Britain, fighter aircraft operating from airfields in Norway, and Fleet Air Arm operations from aircraft carriers. As stated earlier, British Bomber Command was unable to tactically or operationally influence German ground operations in Norway due to range limitations and the vulnerability of its aircraft. Of equal importance was the fact that establishing British fighter aircraft ashore was constrained by the lack of suitable airfields and inadequate lines of communication.[53] Moreover, the strength of that the *Luftwaffe* in Norway had been grossly underestimated by the British Chiefs of Staff.[54] This was especially true not only in terms of the effect on British Fleet operations but also the impact of the German aerial attacks at the port towns of Andalasnes and Namsos.[55]

---

and secondary sources and Hitler's intervention was to have a significant negative impact on German operations as the war progressed.

[51] Ziemke, 94-95. German airborne reinforcements on 14 May initially totaled a mere 66 paratroopers. In the first week of June, however, a parachute battalion and two mountain companies (which were given rudimentary parachute training and dropped at Narvik) were added to General Dietl's command. Of note is the fact that General Dietl had participated in arctic warfare training in the Narvik area during the years before World War II and was well prepared to handle the extreme circumstances his command faced during the campaign.

[52] Churchill, 652. See also Kessler, 25, and Ziemke, 99.

[53] Derry, 231.

[54] Harvey, 170.

[55] Derry, 235. Jack Adams, a soldier in the 24th Guards Brigade during the campaign, provides supporting comments on the psychological impact of German air power on Allied land forces ashore. See Adams, 171-3.

In sharp contrast to the *Luftwaffe*'s ability to influence the campaign, Royal Air Force and the Fleet Air Arm could do little. British developments in "direct air support of naval and land operations were badly neglected during the inter-war years" and this neglect was "glaringly apparent in the Norwegian campaign."[56] Even after World War II, Marshal of the Royal Air Force, Sir John Slessor, wrote, "I cannot call to mind any warlike action against an enemy maritime objective carried out by a Naval aircraft from a carrier, that was not done more often and just as effectively by an RAF aircraft from a base on the shore."[57] Thus, "the dogma of the strategic role of the RAF," coupled with inter-service rivalries between the RAF and the Royal Navy, had "paralyzed the development of maritime-air operations just as it had resulted in the neglect of land-air operations."[58]

In July 1936, the RAF had been reorganized to meet anticipated support requirements for land and sea operations. "The single Command, Air Defence of Great Britain, was abolished and four functional commands were established, Fighter, Coastal, Bomber and Training. Of these Coastal and Bomber played a direct role in the Norwegian campaign, and Fighter Command squadrons were transferred to Norway in support of land operations."[59] Coastal Command was the least prepared of RAF commands at the outbreak of war with nearly all of its 230 aircraft considered obsolete. Accordingly, the mission of attacking the German fleet went to Bomber Command.[60]

Bomber Command also found itself unprepared for war with the majority of its aircraft regarded as obsolete and unable to range the Norwegian Coast. Retired British RAF Air Commodore Maurice Harvey states that the reason behind Bomber Command's inability to range

[56] Harvey, 171.
[57] Sir John Slessor, *The Central Blue*, (Cassell, London, 1956), 191.
[58] Harvey, 183. See also Norman Friedman, Thomas C. Hone and Mark Mandeles, *American and British Aircraft Carrier Development 1919-1941*, (Naval Institute Press, Annapolis, Maryland, 1999), Chapter 4, The Fleet Air Arm: A Failed Revolution?
[59] Harvey, 171.

the coast was that pre-war manufacturers published operational ranges for British aircraft under ideal conditions. Wind, weather, maneuver over the target, or upon attack by enemy aircraft all contributed to substantial reduction in the effective operational radius of the aircraft fielded at the time.[61] The British Chiefs of Staff only realized these shortcomings when the battle was joined.

The RAF's reluctance to pursue deliberate offensive operations in Norway was undoubtedly influenced by heavy Bomber Command losses during the December 1939 raids on the German port city of Wilhelmshaven. These losses clearly demonstrated the vulnerability of Bomber Command aircraft and virtually ended daylight raids by vulnerable British bombers.[62] Of equal significance, Bomber Command also did not possess the navigation equipment, bombsights, or ability to illuminate targets that would have made night bombing a possibility.[63]

In April 1940, anticipating the looming German threat across the channel, the Air Ministry was in the process of building up Fighter Command for defense of the British Isles. This threat, combined with an absence of suitable airfields in Central Norway, precluded the use of high performance modern fighters. Accordingly, it was decided to allocate one RAF squadron of Gloster *Gladiators* to Norway.[64] The *Gladiator,* despite being the last British biplane fighter to enter production, still achieved some success against its adversaries.[65] The *Gladiator's* overall impact was negligible, however, and most were destroyed on the ground, either by German bombers or by the British themselves in order to prevent them from falling into enemy hands.

---

[60] Ibid., 172-3. At the time of the Norwegian campaign, only four British squadrons were equipped with the new American built Lockheed *Hudson,* which had the endurance to reach the Norwegian coast. The *Hudsons* were capable aircraft but could not compete with even mediocre German fighters like the *ME110,* which flew 100 knots faster.
[61] Ibid., 174. Only the RAF's *Whitley* could reach the farthest of airfields in central Norway but it was slow, cumbersome, and had poor defensive armament.
[62] Ibid., 175. The Germans shot down 17 Wellingtons in two raids. German fighter aircraft had quickly learned to attack from above and abeam in order to avoid the Wellington's front and rear turrets, which could not traverse.
[63] Ibid., 175. The British did attempt to bomb Norwegian airfields but realized little success in their efforts. It took nearly three years for Bomber Command to develop its equipment and tactics to operate effectively at night.
[64] Ibid., 180.
[65] Owen Thetford, *Aircraft of the Royal Airforce since 1918,* Seventh Revised Edition, (Putnam, London, 1979), 265.

With Bomber Command and Fighter Command unable to influence the Norwegian campaign for the aforementioned reasons, only Fleet Air Arm could provide the necessary aircraft to protect the British Fleet and lines of communication and forces ashore, as well as provide close air support to ground forces. Unfortunately, although Britain possessed seven aircraft carriers, these ships were outmoded and their aircraft, like the open cockpit biplane *Swordfish*, were more reminiscent of World War I.[66]

Other Fleet Air Arm aircraft included the Gloster *Sea Gladiator* and Blackburn's *Skua* and *Roc*. The *Gladiator* has been described before and the *Sea Gladiator* was little different. The *Skua* differed in that it was the first operational monoplane in the Fleet Air Arm. Originally designed for dive-bombing, it was also expected to serve as a fighter.[67] According to Air Commodore Harvey, it was not effective in either role.[68]

The British were operating aircraft carrier task forces for the first time in Norway in 1940. Additionally, as will be seen again in the Falklands War in 1982, British naval commanders determined that the first task of sea-based aviation was protection of the Fleet. As the aircraft carrier's first priority is self-defense, the result can "often outweigh the residual balance available for offensive action."[69] While the *Gladiators* and *Skuas* did achieve some success, like the sinking of the German Cruiser *Koenigsburg* at Bergen, these aircraft were unable to prevent the *Luftwaffe* from achieving air superiority over Central Norway.[70] The British inability to field sea-based interceptors capable of defeating their land-based German adversaries was also partially the result

[66] Friedman, Hone and Mendelez, 87.
[67] Owen Thetford, *British Naval Aircraft Since 1912,* Sixth Revised Edition, (Naval Institute Press, Annapolis, Maryland 1991), 58.
[68] Ibid., 184.
[69] Harvey, 185.
[70] Friedman, Hone and Mendelez, 88.

of the prewar Royal Navy's conclusion that the fleet would never have enough warning of an air attack and did not develop interceptors in the interwar years.[71]

Employing tactics anticipating those used in the Falklands War of 1982, Vice Admiral Wells, commander of the British carrier task force, positioned his two carriers, *Ark Royal* and *Glorious,* 120 miles off the coast of Norway. Wells was not confident his limited number of embarked aircraft could sufficiently protect his carriers from German land-based aircraft and placed an operational constraint upon his own aircraft to ensure the survivability of his carriers.[72] While Wells succeeded in protecting his carriers from German air attack, he ultimately failed to provide protection for key port facilities, lines of communication, and forces deployed ashore. Thus, in contrast to the role sea-based airpower would play in the Falkland Islands, the British ceded the initiative and air superiority to the Germans, who exploited this advantage in a combined arms role with determined offensive land operations.

Lessons Learned

Professor Olav Riste, perhaps the foremost Norwegian specialist on Operation *Weseruebung*, has synthesized the conclusions of several prominent authors and experts on the Norwegian campaign of 1940. According to Riste, the principle lesson learned from the Norwegian Campaign of 1940 was the undermining of sea-power by airpower.[73] B.H. Liddell Hart disagreed with this conclusion, but conceded that the *Luftwaffe* was the most decisive factor in the German success.[74] According to Adams, Liddell Hart, in his *History of the Second World War*, suggested that German supremacy in the air was more psychological than real as it paralyzed the Allies' countermoves.

---

[71] Ibid., 124. Radar was first placed on British naval vessels in 1938, but the Admiralty failed to develop interceptors, instead relying upon US aircraft as WWII progressed. For an interesting analysis of the organizational and institutional reasons behind the Royal Navy's inability to modernize her naval aviation, see Chapters 5-7, with particular attention to pages 188-99.

[72] Harvey, 185.

[73] Olav Riste, "*Sea Power, Air Power and Wesereubung,*" paper presented to the International Commission for Military History, Washington, D.C. August 1975 (ACTA No.2); quoted from Jack Adams, *The Doomed Expedition, The Campaign in Norway 1940*, (Leo Cooper Ltd., London, 1989), 171.

According to this view, the *Luftwaffe*'s psychological effects far exceed the physical destruction of men and materiel.[75] Regardless of whether the physical or the morale effects were greater, the same effect and end was achieved and the land-based *Luftwaffe* was the decisive factor.

Unlike any previous campaign in military history, *Weseruebung* integrated three armed services and required the cooperation of each as a precondition for success. In Norway, the German Army relied upon the *Kriegsmarine* and the *Luftwaffe* for its very survival as well as its success. This was a significant paradigm shift for the German military because historically Germany was a continental (land-based) power and the Army had always been dominant.

In direct contrast to the British, the Germans task organization for *Weseruebung* demonstrated noteworthy balance and integration, especially during the initial phases of the campaign. The *Kriegsmarine* succeeded in accomplishing its operational tasks despite heavy casualties in the face of vastly superior British naval strength. The *Luftwaffe* proved that sea power alone was no match for airpower and that airpower, under certain conditions, could establish control of the sea and compensate for a weakness in sea power.[76] The *Luftwaffe* contribution throughout the campaign gave the Germans an overwhelming asymmetric advantage against their opponents. The Army achieved all of its initial objectives and inexorably reduced Allied resistance during the remainder of the campaign. The only exception was the battle for Narvik, which ultimately proved inconsequential in the overall operational success of *Weseruebung*.

From the German perspective, *Weseruebung* clearly illustrated operational goals, which were understood by the individual Services. The Services subordinated their inclination to Service parochialism toward joint execution, enabling the Germans to achieve outstanding synergy. This

---

[74] Adams, 171.
[75] Ibid., 171.
[76] Kessler, 3.

resulted in operational and tactical success throughout *Weseruebung*.[77] The German leadership

demonstrated joint planning that appropriately weighed risk through innovation, surprise, and

asymmetric attack. Where British planners dismissed the possibility of German naval operations in the

face of their overwhelming superiority, German joint planners correctly concluded that speed, surprise

and the use of airpower would enable the Germans to seize and occupy Norway in the face of a

superior British naval force. This, despite what General Kessler, the Chief of Staff of *Fliegerkorps X*

during *Weseruebung*, described as "the land mindedness" of the *Luftwaffe* high command and the lack

of training and appropriate weapons to fight the British Navy.[78]

As stated earlier, German operational and tactical use of sea and air transportation to deliver forces

along 1,200 miles of coastline to seize key points supporting the campaign was a noteworthy

achievement for any modern military force. Disregarding secure flanks or a continuous front, the

German armed forces ably balanced the competing requirements of force protection and force

projection, achieving a rapid decision through multi-axis attacks that leveraged speed, tempo, and

mobility to knock out their opponents.[79]

Employing a flexible tri-service German military command structure, German subordinate

commanders had a great deal of independence through mission-type orders. This was in direct

contrast to the Allies' ponderous chain of command, with leaders based in London. The Allies were

unable to match their opponent's decision-action cycle throughout the campaign.[80] Brigadier General

---

[77] MCDP 1-2 *Campaigning* (Department of the Navy, Washington D.C., 1997), 76, defines synergy as the "harmonization of all warfighting functions to accomplish the desired strategic objective in the shortest time possible and with minimal casualties." Although not discussed in detail in this paper, the Germans consistently demonstrated precision and flair in the execution of the six major functions: command and control, maneuver, fires, intelligence, logistics, and force protection.

[78] Kessler, 3. See also Harvey, Chapter XI, "The MasterCard – *Luftwaffe* Operations in Central Norway," 188-200. Harvey concludes that, unlike the British, the German planning staff fully appreciated the key role of airpower and that the capture of airfields and the destruction of the Allied bridgeheads were the decisive actions of the campaign. See also Adams, 173.

[79] MCDP 1-2 *Campaigning*, defines operational tempo as the "pace of events between engagements."

[80] Harvey, 142. Harvey repeats the often-remarked upon conclusion that lessons in command and control are relearned every war. He compares the British campaign in Norway to similar long distance tactical and operational

Richards, one of the senior British general officers who took part in the action in central Norway later commented, "The exercise of any sort of [British] effective command was rendered almost impossible by the complete absence of staff, transport, maps or communications of any sort."[81]

Oversight by the OKW was appropriate given that none of the three Services could supervise the others in a joint effort. This unified command system (an ad-hoc organization in a peripheral campaign) was noteworthy, as the Germans did not have any established joint doctrine or training. But, despite their success (and small number of failures) in Norway, the German armed forces, unlike their British counterparts, failed to build upon these lessons in order to institute joint doctrinal and organizational changes in the German military. German failure to adapt would prove costly in other theaters during the remainder of the war. For the British, adaptation and jointness would enable successful execution of a number of future amphibious operations in Europe.

If the German achievement of surprise and demonstration of the paramount importance of air superiority constitute two outstanding features of the Norwegian campaign, a third many be summed up as the comparative slowness and vacillation which appeared to characterize the British reaction to the German enterprise.[82] The reasons behind this vacillation are two-fold. Mr. C.R. Atlee, a Member of Parliament, stated in the House of Commons that the Government lacked "a settled plan for the vital objective."[83] This criticism was confirmed by Lieutenant General Carton de Wiart and Lieutenant General H.R.S. Massy, the two senior British generals who fought in central Norway, both claiming that intelligence was unsatisfactory and planning guidance from London was "concocted hour to hour."[84]

---

communications difficulties encountered during the Falkland Islands War of 1982. This, despite "modern" satellite communications capabilities available to the British in 1982.

[81] Ibid., 159.
[82] Derry, 235.
[83] Ibid., 236, H. of C. Debates, Vol. 360, Col. 1090.
[84] Ibid., 236.

The British system of command was exercised through the Chief's of Staff but was subject to frequent intervention by the Military Coordination Committee and by the War Cabinet. These interventions were described as sudden and "sometimes seemingly impulsive." Noted author and historian T.K. Derry recommended that the Norwegian campaign should have "been entrusted at the outset to a single Supreme Commander or (more probably) to Commanders-in-Chief from the three Services, functioning through a Combined Headquarters, [then] the situation would have been different." Derry concluded that the formation of a Combined Headquarters would have enabled operations to be conducted "farther away from Ministers [and] would have encouraged the restriction of their intervention to its proper field of grand strategy.[85]

The Norwegian campaign was the first major action involving British forces since World War I. The campaign proved to be a test bed that identified weaknesses throughout the British command structure. Derry describes how the British leadership recognized that their system of command needed restructuring so that it "would contribute to the closest integration of effort throughout the Services."[86] Derry also concluded that this lesson "was duly learnt – perhaps it was the most important lesson learnt in Norway."[87] Putting this into proper perspective it is important to note that this was not only the first major action by the British military since World War I, it was also the first campaign in history requiring the full integration of all three services and the joint nature of the campaign caught Britain by surprise.

Conclusions

Strategically, the impact of the occupation of Norway, *Weseruebung* (*Nord*) and Denmark, *Weseruebung* (*Sud*) is debatable. The Germans could not take advantage of the global reach afforded by Norway's position to occupy Iceland and attack the United States. They were, however, able to

---

[85] Ibid., 238.
[86] Ibid., 238.
[87] Ibid., 238.

secure Germany against Allied attack from Scandinavia until the end of the war. The Germans also obtained additional air and submarine bases for attacks against Britain, which initially forced the RAF to maintain a larger fighter presence in Scotland and later proved invaluable in attacks against Allied supply convoys to Murmansk in the Soviet Union. Perhaps Germany's most important strategic benefit was that, "not only did control of Norway mean that the Germans could ship iron ore to Narvik by train and from Narvik by sea to Germany in winter, but combined with the occupation of Denmark, the occupation of Norway provided a strong position to extort from Sweden almost anything the Third Reich wanted."[88] The German occupation of "Fortress Norway" eventually tied up 300,000 German soldiers. This burden of occupation became onerous as the war progressed as Germany faced the combined manpower of the Soviet Union and the U.S.[89] The Germans also felt the indirect effects of the loss of three cruisers, ten destroyers, and four U-boats, and serious damage to two of her battle cruisers. These losses had important bearing on the German decision not to attempt a cross-channel invasion of Britain in the autumn of 1940.[90] The Germans, in effect, sacrificed a more important strategic goal for the accomplishment of a peripheral objective.

At the operational level, *Weseruebung* established a milestone in the history of joint warfare by clearly demonstrating the effective reach of modern military forces. German planners and commanders sought a rapid decision utilizing speed, surprise, and airpower in a joint, multi-axis attack that obtained tactical decisive action at operational decisive points. While the campaign did not conclude as quickly as German military planners had hoped, it was far less costly in terms of men and materiel than future

---

[88] Weinberg, 119. For an interesting perspective, see Dennis Showalter's "Phony and Hot War 1939-40" in Harold Deutsch and Dennis Showalter's *What If? Strategic Alternatives of World War II* (Emperor's Press, Chicago, USA, 1997), 39-42. See also Derry, Chapter XV, "The Campaign in Retrospect," 229-46.

[89] Harvey, 300-5. Britain obtained control of the major part of Norway's mercantile fleet, the fourth largest n the world. This fleet, particularly the tankers, proved of inestimable value during the darkest days of the Battle of the Atlantic from 1941 onwards.

[90] Churchill, 657. Sir Winston Churchill describes the ruin of the *Kriegsmarine* while desperately grappling with their British counterparts, concluding that it "was thus no factor in the supreme issue of the invasion of Britain."

campaigns in the war.[91] Significantly, it did not cause any material reduction of the forces designated for *Fall Gelb*, and its execution did not interfere with *Fall Gelb's* preparations. In fact, *Weseruebung* provided the Germans with benefits from unintended strategic miscalculation on the part of their Allied opponents in that senior British leadership did not believe that the Germans would execute an attack into the Low Countries while they were engaged in Norway.[92]

Ultimately, mastery of the operational art (service expertise of German commanders), combined with cooperation in all joint considerations, set the conditions for tactical success.[93] These conditions were enhanced by the dominance of German airpower, which assisted in achieving initial surprise, provided unprecedented operational reach, and controlled and constrained the enemy's logistics thereby preventing the introduction of Allied reinforcements or even adequate support for those troops already ashore.[94] With the notable exception of failing to capture the King of Norway and leading members of his government, the campaign's initial objectives were accomplished quickly. Allied forces joined the Norwegians and were able to conduct a number of counterattacks, including the successful recapture of Narvik. They were, however, in the end forced to withdraw from Norway, with the last Norwegian forces surrendering within 60 days of the beginning of the campaign.

---

[91] A lecturer at US Marine Corps Command and Staff College compared Germany's peripheral campaign in Norway with the campaign she later conducted in the Balkans, which set back the invasion of the USSR by several crucial fair weather months.

[92] Field Marshal Lord Alanbrooke, *War Diaries 1939-1945* (Weidenfield and Nicolson, London, 2001). As Chief of the Imperial General Staff (Britain's top Soldier), Field Marshal Alanbrooke's diary entry for 12 April reads, "So far there are only rumors of impending attacks through Belgium and Holland. Personally I feel that this is an unlikely eventuality as the Germans have their hands full with Norway for the present, where they require the bulk of their air force." Author's note – *Weseruebung* served as a "strategic feint" for the Germans as they prepared to execute *Fall Gelb.*

[93] It should also be noted, however, that many of the Norwegian armed forces were conscripts. Also of significance is the fact that many of the British forces that took part in operations in central Norway were Territorials (Reservists activated for the war) with limited training and experience. These troops were initially assigned to execute unopposed landings to occupy undefended objectives. Their mission changed to counter-landings and offensive operations against well trained and equipped German forces using combined arms.

[94] Harvey, 167-8.

Chapter II
Sea-based Air Power and Air Superiority in The Falklands War of 1982

Britain was going to war at the end of a seven and a half thousand miles long logistic pipeline, outside the NATO area, with virtually none of the shore-based air we normally count on, against an enemy of which we knew little, in a part of the world for which we had no specific plan or concept of operations.

- Major-General Sir Jeremy Moore
Commander, Land Forces, Falkland Islands War

With the post-World War II decline of colonialism and empire, Britain's political and military position in the world diminished. Throughout the 1950s and 1960s, many of Britain's colonies gained their independence, and although Britain was still an influential nation on the world stage, it was clear by the 1970s, after a huge military drawdown, that the British would be unable to exert their influence militarily in a manner similar to the past. Indeed, after four centuries of naval dominance, the Royal Navy was under attack from politicians who argued that it was too expensive and not relevant to Britain's current role in the North Atlantic Treaty Organization. "As each patch of red faded from the globe, so too did the need for aircraft carriers, amphibious landing ships and overseas bases." [95]

In 1966, British Secretary of State for Defense, Denis Healy, announced in a Defense Ministry White Paper that he could not foresee any operation that Britain would undertake where aircraft carriers would be needed. Convinced that land-based airpower was the future, Healy remarked that "in the future, aircraft operating from land bases should take over strike-reconnaissance and air defense functions of the carrier.... Airborne early warning aircraft will.... subsequently operate from land bases." [96] Testifying before Parliament, Healy stated, "The fact is that the United States is the only country in the world which plans to maintain a viable carrier force around the world through the 1970s. Neither the USSR or China has carriers or plans

---

[95] Max Hastings and Simon Jenkins, *The Battle for the Falklands*, (Norton, New York, 1983), 10.
[96] Norman Polmar, *Aircraft Carriers; A Graphic History of Carrier Aviation and its Influence on World Events* (Doubleday, New York, 1969), 689.

to have them."[97]

Healy's political assault on the Royal Navy resulted in the resignation of Navy Minister Christopher Mayhew.[98] Although the Royal Navy was forced to abandon "big deck" carriers, there was an acknowledged need for air power in the fleet. Accordingly, naval planners subsequently constructed cheaper, "small deck" carriers to accommodate vertical start take off and landing (V/STOL) aircraft.[99]

In 1981, efforts to remove carriers from the Royal Navy's inventory were again forcefully pursued by Defense Secretary John Nott. Nott believed that "the navy should concentrate on anti-Soviet and anti-submarine defence." Nott believed the function of small deck carriers could be better (and more cheaply) accomplished by destroyers and frigates. As a result, another Navy Minister, Keith Speed, resigned under protest.[100] In their book, *The Battle of the Falklands*, Max Hastings and Simon Jenkins posit, "The whole tenor of Nott's 1981 review -- inspired by the most sustained attack ever mounted by Treasury on defense spending - was to curtail the surface role of [the] navy and reduce its need for costly surface warships."[101] The authors conclude "Nott had finally called the Royal Navy's bluff. Set-piece sea battles, mass convoys, amphibious landings and land-support gunnery -- the textbook maneuvers of the first half of the twentieth century -- had finally been sent to the museum."[102]

In 1982, after 15 years of steady decline in the Royal Navy, and within one year of Nott's criticism of the utility of aircraft carriers, the 300-year feud over possession of Falklands came to a head. On 1 April 1982, shortly after the 150th anniversary of British occupation of the Falkland

---

[97] Ibid., 690.
[98] Hastings, 11.
[99] Bruce Watson and Peter Dunn, *Military Lessons of Falklands War* (Westview Press, Colorado, 1984), 14.
[100] Hastings, 11.
[101] Ibid., 11.
[102] Ibid., 12.

Islands, Argentine Marines and commandos landed, seizing the Islands from the tiny British garrison there.

The British government was now faced with an unexpected war that they were unprepared for. The battlefield was not in Europe or within close reach of land-based aircraft. The enemy was not a Warsaw Pact rival of NATO. Long prepared for the nuclear threat from Eastern Europe and Russia, the British now faced a Western rival with conventional weapons. Perhaps of greatest significance, the British were forced to face this enemy alone, without the help of NATO.

In the 1980s Britain's ability as a former colonial great power to impose its will anywhere around the globe had largely gone the way of the former empire. Although still a major industrialized nation and a nuclear power with strong ties to many nations, including the United States, Britain was no longer a superpower. Why, then, was Britain willing to commit a reduced military capability in defense of territorial possessions of limited economic value so far away from her home islands?

While Britain had shown a willingness to negotiate the status of the Falkland Islands, the British had also, by steadily reducing her armed forces, perhaps unwittingly demonstrated an unwillingness to defend them. Britain's overarching strategic interest was preservation of its national image in maintaining world order. While "she may have been willing to negotiate away her interests, [she] … was not willing to lose them by the force of arms [to] what she considered a third rate power.[103]

Argentina had long desired to "retake" the Malvinas Islands, as the Falklands were referred to by the Argentines. Dr. Robert L. Scheina, a noted author and expert on Latin American sea power, has described Argentina's actions in early 1982 as "heaped upon years of frustration and…the last

[103] Michael J. Nevin "The Falkland Islands - An Example of Operational Art", US Army War College Monograph, Carlisle Barracks, PA April 1986. DTIC Document F3031.5.N48 1986 C.3, 9.

straw."[104] Dr. Scheina concluded that the Argentine government combined a long-standing belief

that the Malvinas belonged to Argentina with the need to establish itself as a regional power in the

South Atlantic and South America. By reclaiming the Malvinas, Argentina meant to assume a

more assertive role in world affairs. Indeed, Dr. Scheina firmly refutes the often-heard opinion

that General Galtieri, the President of Argentina and leader of the military *Junta* government, used

the crisis to divert attention from political and economic problems in Argentina.[105]

In 1982, the Argentine Navy (*Armada Republica Argentina*) possessed an older but well-

balanced and efficient regional fleet.[106] Approximately one-third the size of the British fleet, the

Argentines fielded "one carrier, one cruiser, nine destroyers (two- British Type 42s), five frigates

and four submarines."[107] But the Argentines suffered from "inadequate sealift and amphibious

assault capacity," but compensated with "superior air power, including a small organic naval air

arm."[108]

The Argentines' efforts immediately after seizure of the Falklands can be divided into three

distinct phases. Phase one saw the fleet consolidate its gains, deploy in a show of force, and begin

a training program in the event the British took action in the response to their occupation of the

Islands. This phase ended when the Argentine cruiser, *General Belgrano,* was torpedoed and sunk

outside the maritime exclusion zone, the Argentine submarine, *Santa Fe,* was attacked and ran

aground off South Georgia Island, and the British frigate, *Sheffield,* was sunk by an air launched

Exocet missile.[109] Phase two included an aborted Argentine carrier strike and the ultimate

---

[104] Robert L. Scheina, "The Malvinas Campaign," US Naval Institute *Proceedings*, (Naval Review 1983): 98. Dr. Scheina also states that the Malvinas issue was taught to children in school as an injustice against Argentina by Great Britain.
[105] See also Juan Carlos Murguizur "The South Atlantic Conflict, an Argentinean Point of View," *International Defense Review*, (International Defense Review Volume II 1983): 135-40.
[106] Charles W. Koburger, Jr., *Sea Power in the Falklands* (Preager Publishers, NY 1983), 129.
[107] Ibid., 129.
[108] Ibid., 129.
[109] Ibid., 130.

withdrawal of the Argentine Navy to the shallow waters on the continental shelf.[110] Phase three

found the Argentine surface navy primarily operating along the mainland coast performing anti

submarine warfare (ASW), maritime surveillance, electronic warfare, and coastal defense

missions. The Argentine naval air arm, submarines, and Marines ashore took up the brunt of the

fighting that remained.[111]

British Invasion Force

Lose Invincible and the operation is severely jeopardized. Lose Hermes and the operation is over.
One unlucky torpedo, bomb or missile hit, even a simple but major accident on board, could do it.
- Admiral Sir John Woodward,
Commander South Atlantic Task Groups, Falklands War

Shortly after Argentina's military occupation of the Falklands, Admiral Sandy Woodward was

tasked with leading the 36-warship main body of a British invasion force across 8,000 miles of

ocean to the islands, which lay a mere 400 miles from Argentina. With no airfields nearby capable

of supporting RAF fighters, Woodward was forced to rely on the Royal Navy's two remaining

aircraft carriers, the 23 years old HMS *Hermes* (12 Sea Harriers/18 helicopters) and the recently

commissioned HMS *Invincible* (8 Sea Harriers/15 helicopters).[112]

With only two British aircraft carriers providing air support for the campaign, the total aircraft

match up was 20 British Sea Harriers vs. 223 land-based Argentine fixed-wing jets.[113] This

collection of British aircraft would be responsible for gaining air superiority over the naval task

force while concurrently providing air superiority, air interdiction, and close support for the

landing forces. As Admiral Woodward stated in his memoirs, "We could not, in the face of a two-

hundred-strong enemy air force, put forces ashore anywhere on the islands without air

---

[110] Ibid., 132.
[111] Ibid., 133.
[112] Hastings, 82. The last of Britain's real aircraft carriers, the *Ark Royal*, with her Buccaneers and Phantoms, had been withdrawn from service at the end of 1978.

superiority."[114] Indeed, it was British Air Chief Marshal Beetham's view "that the navy was taking a great risk in committing the task force to action with such slender air cover."[115]

Woodward was a submariner and lacked carrier experience.[116] He did, however, recognize that the carriers were a critical vulnerability stating, "If [the] Argentines knew what they were doing and hit one of my carriers, we would not need a reason to start a war. The war would already be over."[117] Not unlike what the British had faced in Norway in 1940, Woodward's lack of airpower in the Falklands was worsened by his even greater lack of anti-air defense for his amphibious task force and landing forces ashore. Moreover, the primary combat aircraft available for the Falklands campaign -- the Harrier -- was in many respects a less than optimal fighting platform.

The British Sea Harrier possessed a number of limitations, including limited range and an inability to fire radar guided missiles. In addition, the small British carriers could not carry airborne early warning (AEW) aircraft (e.g., E-2s) so the Harriers were blind to incoming threats.[118] In terms of prosecuting a naval campaign away from land-based airpower, many sources concluded "the lack of [Airborne Early Warning and Control] AWACs was the single most critical British deficiency of the War."[119] Had AEW assets been available to Woodward, his limited number of fixed wing aircraft could have been released to perform routine combat air patrols (CAP).[120]

---

[113] Ibid., 219. The 20 Sea Harriers embarked aboard the two British carriers were double the normal complement. Additional pilots were reassigned from instructor and non-flying billets and two were taken from flight training. See also Middlebrook, 72-3.

[114] Woodward, 98.

[115] Middlebrook, 73.

[116] Woodward, 88. Author's note: It is interesting to speculate as to whether the US Navy would ever allow a submariner to command a carrier task force. See also Hastings, 82-3.

[117] Ibid., 108.

[118] The British did have limited AEW capability from *Wessex* helicopters with AEW and Air Search Radars.

[119] Hastings, 142. British requests for US AEW assets were refused due to training time considerations. See also Watson and Dunn, 17.

[120] Ibid., 157. See also Her Majesty's White Paper, *Lessons Learned from the Falklands Campaign*, which describes the lack of AEW as one of the greatest handicaps of the task force. The British were forced to use their submarines as AEW platforms. See also Watson and Dunn, 129. Scrapping their last big deck carrier resulted in the retirement of Gannet AEW aircraft from the Royal Navy's inventory.

It is reasonable to conclude that the British could have reacted to AEW information "to intercept incoming enemy aircraft, either before they came within range of the task force or before they were able to deliver their weapon loads. This would have reduced substantially the threat from sea-skimming missiles; it would also have given the ships of the task force more time to take evasive action and prepare their own active and passive defenses."[121] Approximate closing speeds of attack aircraft and *Exocet* missiles are 480 and 600 knots respectively, with closure rates of 8-10 nautical miles per minute. Under normal conditions, the unassisted radar range of a British Type-21 frigate was approximately 22 miles. Based upon the aforementioned data, when a hostile track was detected, from one minute and forty seconds to two minutes remained for the defenders to react to the aircraft or missile attack. The positive impact on defensive reaction time provided by an AEW aircraft, which could detect an A-4 Skyhawk at a range of 90-100 nautical miles away (22 minutes of warning vice two minutes or less), become obvious. [122]

For technical and doctrinal reasons, the surface combatant air defenses for the British task force was less than adequate. Admiral Woodward also faced the problem of building task force cohesion focused on air defense of the Fleet. Sailing from Portsmouth on 5 April, Woodward described his 36-ship task force in disparaging terms: "...do not imagine that some well-oiled monolith was swinging into action or that any corporate plan had emerged at the early state. We were going to war with virtually none of the shore-based air we normally count on, against an enemy of which we knew little, and in a part of the world for which we had no concept of operations."[123]

The Argentines, however, were aware of British tactics and capabilities. Moreover, their surface combatant inventory included some of the very same technology, including long-range

---

[121] House of Commons Fourth Report from the Defence Committee Session 1986-87, *Implementing the Lessons of the Falklands Campaign*, (Her Majesty's Stationary Office, London, 1987).
[122] Neville Catley, "Airborne Early Warning: A Primary Requirement," <u>Navy International</u>, 88 (January 1983), 34.

Sea Dart missiles from Type-42 destroyers.[124] Significantly, the Argentine Navy was also well aware of the Sea Dart's "one overwhelming weakness: [it was] designed to meet high-flying Russian aircraft [bombers], [and] could not engage targets at low level."[125]

The British did, however, possess the capable Sea Wolf short-range missile.[126] Unfortunately, this system was only present on British Type 22 frigates. This was due to cost saving measures adopted by the British Ministry of Defense (MoD) that had restricted the length of the Type-42 destroyer, enabling it to carry only the Sea Dart system without the Sea Wolf.[127] Thus, the British destroyers and frigates could individually provide only limited anti-aircraft missile defense. More importantly, they could not link their capabilities together for a unified defense of the naval task force and had not trained in this role.[128]

In addition to the shortfall in anti-air defense systems, active defensive counter measures like chaff, against weapons like the *Exocet* missile were present only on Type-22 frigates. Moreover, the British Navy did not posses doctrine to employ these countermeasures. Ultimately, the British task force did not have adequate defensive options against incoming cruise missiles.[129] Nevertheless, Admiral Woodward prepared his force and formulated a strategy for defeating the Argentines.

While the Argentine strategic objective was to assert their sovereignty over the Falkland Islands and increase their standing in the South Atlantic and South America, at the operational

---

[123] Middlebrook, 68.

[124] The Sea Dart's range was 20 nautical miles.

[125] Hastings 116-7.

[126] The Sea Wolf's range was four nautical miles.

[127] House of Commons Fourth Report to the Defense Committee 1986-87, 42. The report concluded that "savings achieved by design modifications should be seen in the context of the capabilities which such modifications may restrict.... It is right that the House remains aware of the penalties that may result from making certain types of saving."

[128] Hastings 152. Later, in the wake of HMS Sheffield sinking, the British changed their tactics by combining destroyers and frigates to form a layered defense against air attacks. Woodward also details adapting Type 22 radars to the tactical situation to compensate for lack of doctrine. See Woodward, 255-6. Also see House of Commons Fourth Report for a thorough after-action review of the performance of all British weapons systems, platforms, etc.

level, their objective was simply to seize and hold the islands. As stated earlier, the British strategic objective was to restore their own sovereignty over the Falkland Islands. Anticipating limited success through diplomatic means, the British government's early determination to use military force supported the accomplishment of operational objectives that included establishment and enforcement of a total exclusion zone (TEZ) and the stated intention to conduct an amphibious operation to land troops in conjunction with a land campaign to seize Port Stanley (assessed as a decisive point).[130]

Operational Planning

As the task force steamed toward the Falklands, Admiral Fieldhouse's staff in London developed the overall strategy for Operation *Corporate,* the British code name for the Falklands Campaign. They envisioned a five-phase operation in which airpower was to play a crucial role.[131] Phase one was the "work-up" or preparation phase. During this phase, the task force and embarked aircraft would conduct maneuvers *en route.*[132] Concurrently, modifications would be made to requisitioned civilian ships taken up from trade (STUFT). Luxury liners like the *Queen Elizabeth II* were converted to troop transports while container ships like the *Atlantic Conveyor* were modified to carry aircraft. British combat aircraft were also modified, including the RAF GR3 Harrier (ground attack version), which was modified within a week and joined the fleet with the war ongoing.[133] The GR3 was modified to carry the AIM-9 "Sidewinder" missile and had an improvised Inertial Navigation System (INS) installed to allow for operations afloat. Six RAF

---

[129] Hastings 116. The House of Commons Fourth Report cites numerous corrective measures related to this critical deficiency.

[130] Harry D. Train II. "An Analysis of the Falklands/Malvinas Islands Campaign." *Naval War College Review,* (Winter 1988, 33-50): 43. Considering these operational objectives, Admiral Train commented, "As the seat of government, the center of population, and the location of the principal seaport and airfield, Port Stanley was the key to the campaign."

[131] General Jeremy Moore and Admiral Sandy Woodward "The Falklands Experience" Lecture given at the RUSI on 20 October 1982, *RUSI/Journal of the Royal United Services Institute for Defense Studies* 128, No. 1 (March 1983), 25-32.

[132] Moore and Woodward, 27.

Vulcan bombers were also converted to serve as tankers and some Nimrod maritime patrol aircraft were modified to carry AIM-9 and Harpoon missiles.[134] The exceptional circumstances of the campaign challenged the British government to harness military and civilian industrial cooperation as evidenced by the rapid modifications made to military and civilian equipment. The speed with which these modifications were accomplished is a testament to the national sense of urgency demonstrated by the crisis.[135]

Phase Two consisted of a blockade of the Falklands.[136] The British campaign plan identified the Argentine forces occupying the Falkland Islands as Argentina's operational center of gravity. The Argentine forces on the islands relied on logistical support from the mainland and therefore could be isolated. To achieve this end, the British declared a 200-mile or total maritime exclusion zone (MEZ) around the Falklands. Upon arrival of the Sea Harrier-equipped task force, the zone was vertically expanded as a TEZ. The TEZ became effective on 30 April and applied to all Argentine aircraft or shipping. Woodward's assessment of the blockade, and the British ability to stop determined Argentine attacks was summed up by his speculative comment, "What if they made a really determined effort of, say, fifty aircraft in a major strike? What if they are prepared to lose twenty or thirty aircraft in an all-out attempt to sink one of our carriers?"[137]

During this phase, the British also planned to recapture South Georgia Island as a major demonstration. The recapture of South Georgia Island was intended to restore British territory, provide a deep port at a safe distance for the fleet, provide a morale boost for British forces and the people at home which, it was hoped, would also would demoralize the Argentines, and finally illustrate British resolve.

---

[133] Her Majesty's White Paper, "Lessons Learned in the Falklands" (Her Majesty's Stationary Office, London), 19.
[134] Ibid., 20.
[135] Author's note: the campaign demonstrated the value of a broad based national defense industry and the benefits of an in-house research capability. See Her Majesty's White Paper, 24.
[136] Moore and Woodward, 27.

In the event that the blockade and diplomatic measures were unsuccessful, phase three envisioned a major amphibious assault, immediately followed by phase four, which would be sustained operations ashore to defeat the Argentine forces. Phase five was regarded as mopping up and police actions, with the Falkland Islands returned to British rule upon the surrender of the Argentine forces.[138]

Role and Effectiveness of Airpower

Although there were no Coral Sea or Midway carrier battles around the Falklands, carriers played a crucial role. During the Argentine invasion, the Argentine Navy's *Veinticinco de Mayo* provided carrier-based air support to the 2,000 troops that invaded the Falklands.[139] While these aircraft were not needed in the attack, they could have brought considerable firepower against the island's civilians and small Royal Marine detachment if required.

Land-based Argentine strike aircraft were the principal threat to the British task force and amphibious operations. As stated earlier, in order to protect the task force, the role of defensive counter-air was of primary importance in British carrier employment considerations.

The Royal Navy began air attacks against military targets in the Falklands with carrier-based Sea Harriers on 1 May. [140] Although the Harrier is a V/STOL aircraft, the trade off is that its payload is less than that of a comparable conventional aircraft. The British did not possess any other fixed-wing aircraft that could operate from their short V/STOL carriers. But the Sea Harriers were also augmented later in the conflict by RAF GR.3 Harriers to provide improved strike capability in support of ground operations.

During the conflict, the Sea Harriers flew 1,100 air defense sorties and Sea Harriers and

---

[137] Woodward, 99.
[138] Moore and Woodward, 27.
[139] Watson and Dunn, 16.
[140] Department of the Navy Summary Report, *Lessons of the Falklands* (Department of the Navy, Washington, D.C. February 1983), 27.

Harrier GR.3 aircraft together flew 215 ground attack sorties.[141] The successful performance

of the British during the conflict can be attributed to the excellence of Royal Navy

and RAF pilots and maintenance personnel, combined with the simplicity and soundness of the

Harrier design. These accomplishments are particularly noteworthy given the adverse weather

conditions in the South Atlantic. The V/STOL Harrier, however, was not designed to provide air

superiority and largely failed to protect British forces from attack.[142]

Overall, the British Harriers demonstrated a very high readiness rate (approximately 90%),

good all-weather flying capability, a long ferry range with the aid of in-flight refueling, low

maintenance requirements, and a low combat loss rate. However, they did experience a rather high

accident rate with five operational non-malfunction accidents over six weeks of combat. [143]

Skillful employment of available aircraft and positioning of the task force enabled the British to

achieve local air superiority for short periods. The Sea Harriers were used as the first line of a

"defense in depth" with the second line assumed by guided missile destroyers and frigates

positioned along anticipated Argentine attack axes. Accordingly, Admiral Woodward positioned

his carriers at the eastern extreme of the TEZ, effectively out of reach of Argentine strike aircraft.

This posed the disadvantage, however, of limiting the Harrier's on-station time for CAP. In order

to provide three CAPs of two aircraft each, 18 aircraft had to be airborne simultaneously, with six

on-station, six *en route*, and six returning to base.[144] It is easy to conclude that it was only due to

the Harrier's excellent serviceability record and the superb work of the maintenance crews that the

British were able to maintain CAP with the limited number of platforms available.

---

[141] Ibid., 27. The Harriers could carry up to three 1,000lb bombs or other air-to-ground weapons in a close air support role.
[142] Ibid., 27.
[143] Department of the Navy Summary Report, *Lessons of the Falklands*, 27. See also Burden, *Falklands: The Air War* (Arms and Armour Press, London 1986), 189-90. See also House of Commons Fourth Report, 59-61.
[144] Burden, 190.

British offensive counter air efforts were conducted in support of the primary mission of defending the task force. The Royal Navy pilots were not trained in air-to-air engagements and their aircraft were ill-equipped to that end. Indeed, the Sea Harrier was designed for fleet defense and maritime attack missions and its radars were not ideal for ground attack, nor were the British pilots trained in that role.[145] With the addition of RAF GR.3 Harriers, the number of close air support missions in support of landing forces ashore increased dramatically. The GR.3 aircraft was designed for this role and the RAF pilots were trained to execute these missions. On 5 June, GR.3 Harriers began to operate ashore from a forward operating base (FOB) at San Carlos, improving response, loiter and turnaround times to ground commanders.[146]

Tactical reconnaissance was limited and usually conducted only in conjunction with ground attack missions. The British did, however, execute long-range operational bombing missions. These missions began on 1 May, using the Vulcan B.2 bomber. These aircraft had been in the process of being phased out in 1982, and were used in anger in the Falklands for the first time in 25 years of service.[147] After being refitted with refueling probes, the Vulcans sortied from Ascension Island, bombed Port Stanley airfield in the Falklands, and returned to Ascension. The missions were 16 hours in duration.[148] In addition to any physical destruction accomplished by dropped ordnance, the missions provided around-the-clock harassment of Argentine forces on the Islands and clearly demonstrated to Argentine leadership that the Argentine mainland was vulnerable to British bombers.[149] The British hoped that this threat would require the Argentines to maintain fighter interceptor aircraft on standby in the event that the British elected to strike the

---

[145] Ibid., 190.
[146] Moore and Woodward, 31.
[147] Burden, 363.
[148] Ibid., 365.
[149] Author's note: British airfield attack missions against Port Stanley did not prevent subsequent Argentine use. The introduction and subsequent use of the JP233 cratering and area denial weapon, used to great effect against Iraqi airfields in Operation Desert Storm in 1991, has since rectified this deficiency.

mainland.[150] Ironically, the Argentine government had inquired about purchasing Vulcans prior to the war as the British preceded with decommissioning the bomber fleet.[151]

The approximately 200 British rotary wing aircraft played a crucial role in the campaign, flying over 20,000 hours. "Helicopters were used in almost every conceivable role; to load ships before departure and en route; to transfer stores during the passage south and in area; for ASW missions; for anti-surface ship warfare; for re-supply after landing and movement of stores and ammunition; for tactical movement of troops; for reconnaissance and command purposes; for search and rescue; and for casualty evacuation."[152] Rotary wing assault support and logistics was significantly handicapped by the loss of the *Atlantic Conveyor* and seven of her eight embarked CH-47 Chinook transport helicopters. As a result, Sea King helicopters carried out most of the ASW and the bulk of the logistics and medevac duties.[153]

The Argentine Air Force and Navy began air attacks against British naval forces on 1 May.[154] The Argentines flew approximately 300 sorties against British shipping over the six-week conflict. These sorties were mostly conducted with 500lb and 1,000lb conventional bombs, but also included rocket and strafing attacks and five Exocet missiles launched by Super Entendard aircraft as well as unguided rockets fired by MB.339 aircraft.[155]

Argentine Mirage, Skyhawk, and Super Entendard aircraft flew determined strikes against the British task force and amphibious landings but their efforts were extremely limited by the range of

---

[150] Woodward, 99. It took 11 Victor tanker aircraft to enable one Vulcan bomber to reach and overfly the Falklands for five minutes in a 7,860 mile round trip from Ascension Island.

[151] John Laffin, *Fight for the Falklands* (St. Martin's Press, NY, 1982), 95. "Acquisition of such a long range bomber force would have made Argentina the only Latin American nation with a strategic bombing capability and it would've changed the entire complexion of the balance of forces in the South Atlantic."

[152] House of Commons Fourth Report, 61. See also Burden, 240. As an example of the extensive ASW operations conducted by British pilots, 820 Naval Air Squadron (nine Sea Kings) flew 4,700 hours, which averaged 321 hours per pilot during the conflict.

[153] House of Commons Fourth Report, 62. Sorties of between six and seven hours were common, and some Sea King pilots flew up to ten hours per day. In the landing at San Carlos, some 20,000 tons of equipment were landed, and the equivalent of over 9,000 troops in one day!

[154] Department of the Navy Summary Report, *Lessons of the Falklands*, 25.

the target area from their airfields. This restricted their ability to maneuver and limited their available time on target to engage in strikes or air-to-air combat. The Argentine Air Force did possess a limited in-flight refueling capability with two KC-130 Hercules aircraft plus Skyhawk aircraft using "the buddy store" scheme. But only the Argentine Navy Super Entendard, and the Air Force and Navy Skyhawks could be refueled in flight.[156]

Of great significance was the fact that many of the Argentine bombs did not explode because fuses had not been correctly set for low-level delivery. In addition, the Argentines lacked a number of modern electronic aids and they were only able to mount attacks in daylight and in reasonable weather.[157] While the British did not consider the Argentine Air Force to be sophisticated by modern standards and destroyed nearly one hundred Argentine aircraft, they acknowledged the fact that the Argentines did sink six British ships and damaged many others.[158]

Thus, in the end, Admiral Woodward had only Sea Harriers with which to gain and maintain local air superiority and without the aid of early warning radar. This, coupled with inadequate shipboard air defense systems against low flying aircraft and an inability to link the shipboard defenses that he did possess, created a very challenging environment in which to operate against his Argentine adversaries. Thousands of miles from the nearest land-based airfield, Admiral Woodward would have been quite unable to carry out the British recapture of the Falklands without the use of sea-based aviation.

---

[155] Ibid., 26-7.
[156] Ibid., 27.
[157] Scheina, 114.
[158] House of Commons Fourth Report, 62. The Report lauds the fact that only one merchant ship was sunk and two Royal Fleet Auxiliaries damaged by air attack out of the 77 ships in the two categories which participated and required protection in "the most adverse circumstances." Most of the Argentine aircraft were destroyed during air attacks.

Lessons Learned

While British casualties were significant, involving several ships sunk and aircraft lost, and while the British never established complete air superiority, they were able to project impressive sea-based airpower, and this was the decisive factor in the British victory.

-Bruce Watson and Peter Dunn,
Military Lessons of the Falkland Islands War:
Views from the United States

The limited number of Sea Harriers, combined with a lack of AEW radars, created a formidable challenge for British commanders. With no possibility of air supremacy or even sustained air superiority, the British faced an environment where command of the air remained contested throughout the conflict. Of operational and tactical significance, the British government intentionally limited the Falklands War, ruling out strikes against targets on the Argentine mainland. This deliberate restraint allowed a distant sanctuary for the Argentine Air Force, thereby adding significant risk to the campaign.[159]

Admiral Woodward adapted and creatively applied his weapon systems in "defense in depth," with Harriers, missile destroyers/frigates, and shipboard close-in weapons systems layered along known enemy axes of approach. As feared by the Argentines, the British soon discovered that, because of the combined limitations of fuel, safe altitude, and topography, Argentine aircraft were limited to using only three tracks across West Falkland, and the Sea Harriers were vectored accordingly.[160] These tactics, combined with the placement of his carriers at eastern extreme of the TEZ, reduced the overall vulnerability of the task force to Argentine air attack. The trade-off was to put the Harriers at their extreme range over the Falklands, reducing British CAP to 20 minutes on-station.[161]

---

[159] Stewart Menaul, "The Falklands Campaign: A War of Yesterday?" *Strategic Review*, 89.
[160] Burden, 23.
[161] Department of the Navy Lessons of the Falklands Summary Report (Office of Program Appraisal Washington, D.C., Feb 1983. Author's note: Admiral Woodward questioned his own decision to layer the Type 42 Destroyers and Type 22 Frigates after the loss of the HMS *Coventry*. See Woodward, 289 and 292.

While the British could not have succeeded in the Falklands campaign without airpower, it took the collective leverage and synergy of air, sea and land forces to achieve success. Like the Norwegian campaign of 1940, all the British armed services were dependent upon one another for success in the conflict. As stated earlier, the most challenging mission facing the Royal Navy was providing air cover for the task force and the amphibious assault -- in particular, the vulnerable beachhead during landing operations. That the British never achieved air supremacy was no surprise, but their ability to maintain air parity throughout was due in part to the Argentines' failure to concentrate their airpower at decisive points for decisive action. It is important to note that whenever they tried to overwhelm British defenses, the Argentines experienced success. Despite these successes, the Argentines nevertheless lost the air-to-air battle.

The performance of the Harriers was one of the surprises of the Falklands War. In addition to their unique maneuvering characteristics, the Harriers possessed a trump card in the AIM-9L Sidewinder. These missiles had significantly greater capability over their predecessors used in the Vietnam War, possessing an "all aspect" attack capability, enabling them to fire at approaching targets as opposed to only from the rear.[162] Owing in large measure to this advantage, Argentine aircraft failed to shoot down a single Harrier during the conflict. The Harriers, on the other hand, were responsible for shooting down nearly one half of the Argentine aircraft lost during the conflict (approximately 30 to 40 aircraft).[163]

The Argentines appear to have grossly neglected the possibility of improving the runway at Port Stanley in order to station combat aircraft on the Falklands. "According to one Argentine source, aluminum runway material was "available" but there was no time to ship it. This source claimed that the size and weight of the materials required a large vessel which was not available."[164] It is

---

[162] Watson and Dunn, 42. See also, *Her Majesty's White Paper*, 9.
[163] Koburger, 70. See also Burden, 190.
[164] Scheina, 105.

conceivable that planning errors like this might be directly attributed to an Argentine miscalculation of British resolve. "Failure to fully exploit Port Stanley's short but hard-surface civilian airfield....could have been the key to victory, much as Henderson Field on Guadalcanal in 1942, impervious to torpedoes and almost so to bombs."[165] The Argentines had three weeks between their invasion and the arrival of the British fleet to improve Port Stanley airfield. This would have added 400 additional miles of land-based reach from Port Stanley against any British maritime operations. Argentina's willingness to remain on the strategic and operational defensive was a grave error that passed the initiative to their adversary.

While the first priority of British airpower was to protect the fleet, the second objective was to provide CAS for the Army and Royal Marines when they went ashore.[166] Conversely, the Argentines planned to " to damage the British landing force during the landing when their freedom of movement was limited."[167] Ultimately, this translated into a simple Argentine strategy of destroying British ships. Because Argentine aircraft were operating at the extreme end of their combat envelope they did not have the fuel to search for troops ashore, even considering the lack of vegetation on the islands. The best way to stop the invasion, it was reasoned, was to attack embarked troops before they went ashore or to make the cost to Britain so high in ship losses that Whitehall would opt to withdraw or negotiate.[168]

Foul weather and low ceilings covered British landing operations on 21 May. Yet, when the weather cleared, and the Argentines attacked in mass, they lost 15 aircraft (12 to Harriers). Conversely, the British lost just one Harrier to ground fire. Argentine aircraft successfully destroyed the frigate HMS *Ardent* but many Argentine bombs hit British ships and failed to detonate. Of greatest significance is the fact that none of the British amphibious assault or logistic

---

[165] Koburger, 138.
[166] Watson and Dunn, 43.
[167] Train, 39.

ships were ever hit. According to Admiral Woodward, the "loss of escorts was biggest stroke of luck and [the] enemy's single biggest mistake."[169]

British offensive counter-air and the lack of strategic bombing were disappointments. Not only was the British bombing of Argentine occupied airfields ineffective, the British Harriers only destroyed nine Argentine aircraft on the ground. These figures are in stark contrast to the efforts of the Special Air Services (SAS) Commandos who destroyed 11 aircraft in a daring raid at Pebble Island. That Port Stanley's airfield remained open throughout the campaign is arguably one of the biggest British failures during the conflict.

The importance of secure air bases in the theater of operations is paramount. This lesson was immediately apparent to the British maritime and land forces that participated in the Norwegian Campaign of 1940. In the Falkland Islands War, the Argentine Air Force was forced to fight at the extreme limit of its operational range. One hundred miles in either direction would have significantly affected the outcome of the conflict. Were the Falklands one hundred miles further out in the Atlantic, the Argentine Air Force would have had no impact on the war. Conversely, were the Islands 100 miles closer, the Argentine Air Force probably would have established air superiority and forced the British to adapt accordingly, with all the risks that would have entailed.

As Dr. James L. George, a noted author, historian and national security affairs specialist commented, "Clearly, one of the bright spots was the survivability of the British task force and amphibious assault landing. This survivability can be directly related to the protection of sea-based air power. While the Argentine leadership placed the heaviest load on air force, British air power made its heaviest contribution as part of an integrated combat effort. Air power, one essential element of joint force, played key role in determining victory. There is one general lesson that pertained to the Falklands conflict just as it pertained to any battle during World War II:

---

[168] Watson and Dunn, 44.

air superiority is an absolute necessity. If the Royal Navy had lacked even the limited V/STOL

carriers the British would not have been able to attack the islands. Conversely, had the Argentines

stationed some of their planes on the Falklands, they might have won. Air superiority is as vital

today as it was during any battle of World War II."[170] In the end, "Great Britain won by

destroying enough of the Argentine Air Force and Navy that further Argentine operations on the

Falkland Islands were impossible. The Argentineans [sic] were beaten even though large ground

forces were available on the mainland."[171]

---

[169] Woodward,108.

[170] Watson and Dunn, 17.

Chapter III
Conclusion: Lessons Learned

"If you want to go anywhere in modern war, in the air, on the sea, on land, you must have command of the air."

Fleet Admiral William F."Bull" Halsey

The operational implications of the Norwegian campaign and the Falklands War for the U.S. Navy and Marine Corps team are readily apparent to anyone who has observed and understands that naval forward presence is an essential element of United States' military strategy and foreign policy. The United States is and will remain a maritime nation and "it is therefore imperative that our naval forces remain preeminent as they advance into the 21st Century."[172]

The National Military Strategy of the United States of America, September 1997, articulates four strategic concepts: strategic agility, overseas presence, power projection, and decisive force. In consonance with our National Military Strategy, *Joint Vision 2010*, describes a new operational framework based upon four concepts: dominant maneuver, precision engagement, focused logistics, and full dimensional protection. As stated in the US Navy's visionary document, *Decisive Power from the Sea*: "The implementation of these strategic concepts falls, to a large extent, on the Nation's naval forces."[173]

The above statements reflect the United States' reliance on the use of the sea. To this end, power projection from the sea will require sea-based airpower. Naval aviation exists to provide air superiority for naval fleets and sealift assets, which will be essential to U.S. success in projecting power abroad. Those who would challenge our Navy's ability to exercise positive sea control and to execute amphibious operations will adopt a strategy of sea denial, attacking our sources of strength. Adversaries will target our aircraft carriers and military sealift vessels through the use of

---

[171] John A. Warden III, *The Air Campaign.* (, Paergamon-Brassey's New York, 1989), 113. Author's note: for a thorough review of the statistics related to aircraft losses on both sides, see Burden, 26-7.
[172] Naval Amphibious Warfare Plan (NAWP), *Decisive Power from the Sea*, Dept of the Navy, Wash, D.C. 1999, 6.
[173] Ibid., 6.

mines, shore-based guided missiles, land-based aircraft, fast attack craft, and submarines. Additionally, domestic political sensitivity to the possible large loss of life associated with the loss of a large capital ship compounds the difficulties we could face and may significantly reduce the capability gap between the U.S. and its potential adversaries.

As a maritime nation and the world's sole remaining superpower, the United States can expect to be engaged in peripheral conflicts not unlike what the British faced in 1940 in Norway and again in the Falklands in 1982. Much as the British discovered in the case studies reviewed in this essay, sea-based aviation capabilities will be required to ensure access to the littorals, especially in peripheral conflicts where we are unable to land-base tactical aircraft. As Admiral Woodward commented, "Due to the Argentine's ability to interdict British forces with tactical aircraft from the mainland, carrier airpower was the key to the entire operation: without it, there was no way we could have reclaimed the Falklands."[174] This remains as true today for US planners as it did when Admiral Woodward led the British task force in 1982.

With the present size of our naval amphibious shipping at a historic low since World War II, it is debatable whether we can realistically expect to conduct opposed large-scale amphibious operations. Moreover, the worldwide proliferation of missile technology ensures that conducting naval campaigns in a high-threat anti-ship environment is a reality we must be prepared to deal with wherever our Navy and Marine Corps team is committed. As the British learned the hard way in the two case studies presented in this paper, without air superiority ground forces ashore and vessels of the amphibious task force are at extreme risk to land-based or sea-based enemy aircraft.

---

[174] Woodward, 108.

Further, and arguably just as important, is the requirement for sealift sustainment of any land campaign. Historically, 90-95% of all military supplies are moved by sealift.[175] A modern naval warfare example of sealift vulnerability was aptly demonstrated by the sinking of the *Atlantic Conveyor* in the Falklands War. Thus, a permissive air environment will also be required for the introduction of Maritime Prepositioning Forces and sustainment from the sea.

The U.S. has reduced overseas bases and forward deployed forces in the Post-Cold War period. The aforementioned reductions have lengthened crisis response times and placed greater emphasis on maritime options for the President of the United States and the Secretary of Defense. In effect, "opportunities and challenges in the world's littoral regions will increase America's reliance on the continuous forward presence and sustainable maritime power projection of Naval expeditionary forces."[176]

With a likely scenario of littoral-based operations and the potential non-availability of friendly land-based airfields, the Joint Force Commander will necessarily employ sea-based airpower in the early stages of any campaign to establish battlespace dominance. Integrating land, and air forces, he will synchronize the wide range of capabilities available into full dimensional operations against the enemy. A balanced approach to any peripheral conflict will thus include sea control as a crucial element, heavily dependent on airpower for surface cover and power projection. Aircraft carriers, including their contribution to ship survivability, AEW, and the other functions of aviation, will be crucial to the successful conduct of any campaign. The presence of an enemy land-based aircraft threat can only add greater emphasis to this requirement.

---

[175] NDP 1, 24.

De'nouement

There is also the inescapable truth that the Argentine commanders failed inexplicably to realize that if they had hit Hermes, the British would've been finished. They never really came after the one target that would surely have given them victory.

-Admiral Sir John Woodward,

Commander South Atlantic Task Groups, Falklands War

Aircraft carriers provide fleet support, destroy enemy fleets, protect merchant shipping, destroy enemy merchant shipping, and project firepower in support of amphibious assaults and strategic attacks. As threats in the form of space based sensors, advances in cruise missile technology, and land-based aircraft with increased weapons ranges and lethality continue to proliferate around the globe, our aircraft carriers will remain lucrative targets. As we have seen in the two case studies presented in this paper, a credible threat from land-based airpower will force fleet commanders to execute their primary mission of protecting the fleet. This can result in a reduction or loss of offensive air support or close air support for forces ashore. With the success or failure of an amphibious operation linked to air superiority, the implications of withdrawing aircraft carrier support are readily apparent.

The British experiences highlighted in this essay can also be compared to the Japanese experience at the Battle of the Coral Sea in May of 1942, and our own experience at Guadalcanal in August of that same year. At the conclusion of the Battle of the Coral Sea, with a credible U.S. threat to their carriers, the Japanese abandoned their planned invasion of Port Moresby, New Guinea. At Guadalcanal, the U.S. Navy withdrew its ships and left the landing force ashore without naval support until the Americans were able to win the Battle of the Solomon Islands weeks later.

Without air superiority, all ships become targets in times of war against a determined enemy. These lessons were learned in World War II and relearned in the Falklands decades later. Today,

---

[176] *Marine Corps Strategy 21*, Department of the Navy, United States Marine Corps, November 2000, 3.

and in the foreseeable future, air superiority for ships at sea can only be effectively provided by large aircraft carriers. When planning the conduct of naval campaigns, strategists intending to employ amphibious forces ashore must heed historic precedent for withdrawal of aircraft carriers in the face of a credible threat from land-based airpower.

It is not within the scope of this paper to provide answers to the challenges leaders will face when employing aircraft carriers in a high anti-ship threat environment. Nonetheless, these challenges will be faced by our Navy and Marine Corps team in pursuit of future Expeditionary Maneuver Warfare campaigns. It is the hope of this author that our leaders will not ignore the lessons of history presented in this essay's two case studies.

Bibliography

Adams, Jack, *The Doomed Expedition, The Campaign in Norway 1940*, Leo Cooper Ltd., London, 1989.

Alanbrooke, Field Marshal Lord, *War Diaries 1939-1945*, Weidenfield and Nicolson, London, 2001.

Armitage, M.J. and Mason, R.A., *Air Power in the Nuclear Age*, University of Illinois Press, Chicago, 1983.

Assman, Kurt, (former Vice Admiral German Navy W.W. II), *"The German Campaign in Norway, Origin of the plan, execution of the operation, and measures against Allied counter-attack."* Tactical and Staff Duties Division (Foreign Documents Section), Naval Staff Admiralty. Her Majesty's Stationary Office. London, 1948.

Bekker, Cajus, *The Luftwaffe War Diaries*, Ballantine Books Inc., New York, 1969.

Brown, Eric, *Wings of the Navy, Flying Allied Carrier Aircraft of World War Two*, Naval Institute Press, Annapolis, Maryland, 1987.

Burden, Rodney, *Falklands The Air War*, Arms and Armour Press, London, 1986.

Catley, Neville, *"Airborne Early Warning: A Primary Requirement,"* Navy International, 88 (January 1983).

Churchill, Sir Winston S., *The Gathering Storm*, Houghton Mifflin Company, Boston, 1948.

Classen, Adam R., *Hitler's Northern War, The Luftwaffe's Ill-Fated Campaign, 1940-1945*, University Press of Kansas, 2001.

Deutsch, Harold C. and Showalter, Dennis E., *What If? Strategic Alternatives of World War II,* The Emperor's Press, Chicago, USA. 1997.

Derry, T.K., *The Campaign in Norway*, Her Majesty's Stationary Office, London, 1952.

Ethell, Jeffery and Price, Alfred, *Air War South Atlantic*, McMillan, New York, 1983.

Friedman, Norman, and Hone, Thomas C., and Mandeles, Mark D., *American and British Aircraft Carrier Development 1919-1941*, Naval Institute Press, Annapolis, MD 1999.

Girrier, Robert Paul, *Lessons from the Falklands War: Implications for US Naval Policy in NATO Northern Flank Maritime Operations*, University Microfilms International, Ann Arbor Michigan, 1991.

Green, William, *Warplanes of the Third Reich*, Doubleday and Company, Great Britain, 1972.

Gunston, William, *An Illustrated Guide to Bombers of World War Two*, Salamander Books, London, UK, 1980.

Haikio, Martti, *Scandinavia during the Second World War, The Race for Northern Europe, September 1939-June 1940,* University of Minnesota Press Minneapolis, 1983.

Harvey, Maurice*, Scandanavian Misadventure, the Campaign in Norway 1940,* Tunbridge Wells, United Kingdom,. 1990.

Hastings, Max and Jenkins, Simon, *The Battle for the Falklands*, Norton, New York, 1983.

Her Majesty's White Paper, *Lessons Learned in the Falklands*, Her Majesty's Stationary Office, London, 1982.

House of Commons Fourth Report from the Defence Committee Session 1986-87, *Implementing the Lessons of the Falklands Campaign*, Her Majesty's Stationary Office, London, 1987.

Humble, Richard, *Aircraft Carriers*, Winchmore Publishing Ltd, Hadley Wood, England, 1982.

Kersaudy, Francois, *Norway 1940*, Collins and Sons, London, 1990.

Kessler, Ulrich O. (former General der Flieger German Air Force W.W. II and Chief of Staff, Fliegercorps X during Operation *Weseruebung*) *The Role of the Luftwaffe in the Campaign in Norway 1940*, Army Historical Center Monograph, 1954.

Koburger, Jr., Charles W., *Sea Power in the Falklands*, Preager Publishers, New York, 1983.

Laffin, John, *Fight for the Falklands*, St. Martin's Press, NY, 1982.

Marine Corps Doctinal Publication 1-2, *Campaigning*, Department of the Navy, Washington D.C., 1997.

Menaul, Stewart, *"The Falklands Campaign: A War of Yesterday?"* Strategic Review, 10 (Fall 1982).

Polmar, Norman, *Aircraft Carriers; A Graphic History of Carrier Aviation and its Influence on World Event*s, Doubleday, New York, 1969.

Slessor, Sir John, *Central Blue*, Cassell, London, 1956.

Smith, John Richard and Kay, Antony, *German Aircraft of the Second World War*, Putnam, London, 1972.

Taylor, Telford, *The March of Conquest*, Nautical and Aviation Publishing Company of America, New York, 1991.

Thetford, Owen, *British Naval Aircraft Since 1912, Sixth Revised Edition*, Naval Institute Press, Annapolis, Maryland 1991.

Train, Harry D. II. "*An Analysis of the Falklands/Malvinas Islands Campaign.*" <u>Naval War College Review</u>, Winter 1988, 33-50.

Warden III, John A., *The Air Campaign*, Pergamon-Brassey's New York, 1989.

Watson, Bruce and Dunn, Peter, *Military Lessons of the Falklands Islands War: Views from the United States*, Westview Press, Boulder Colorado, 1984.

Weinberg, Gerhard, *A World at Arms*, Cambridge University Press, United Kingdom, 1997.

Westphal, Siegfried, *The German Army in the West*, Cassell and Company, London, 1951.

Woodward, Sandy, *One Hundred Days, The Memoirs of the Falklands Battle Group Commander*, HarperCollins, London, 1992.

Ziemke, Earl Frederick. *The German Northern Theater of Operations,* Washington D.C., Center of Military History, 1989.